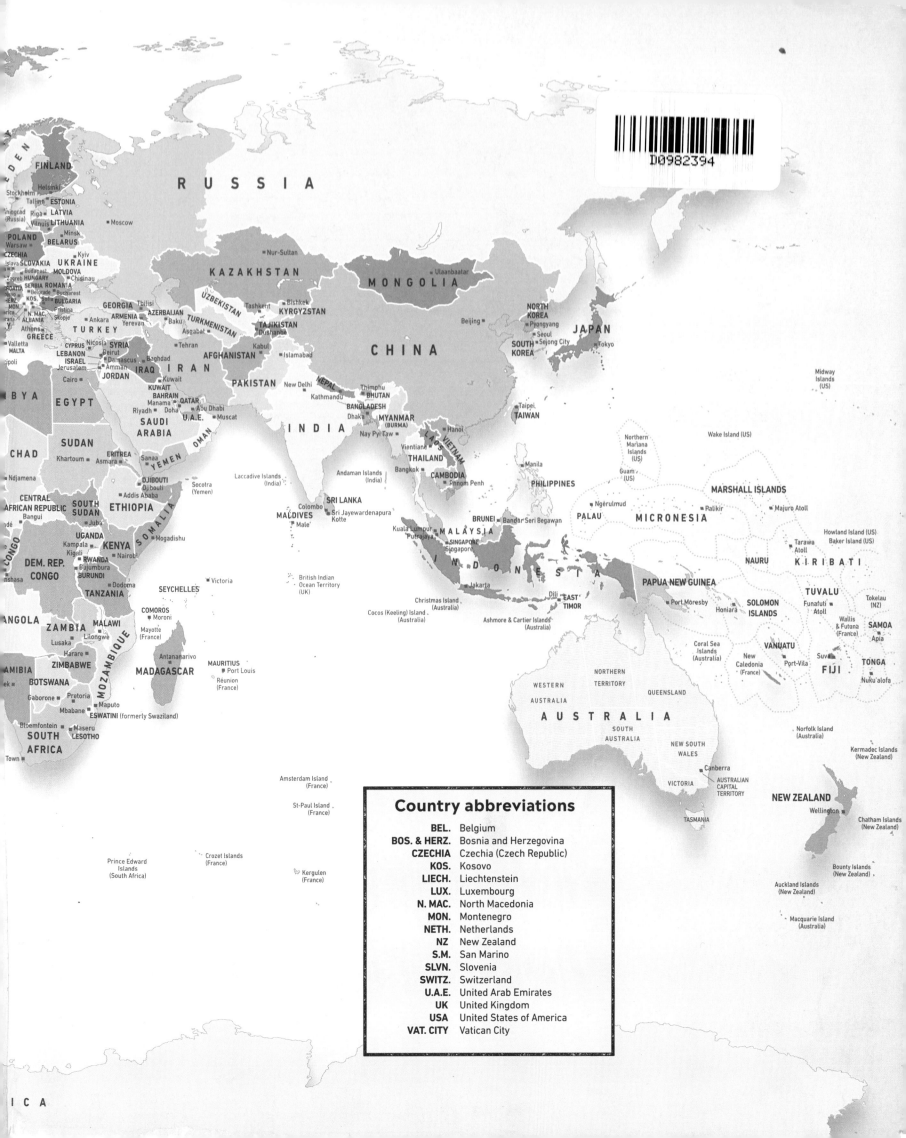

Country abbreviations

BEL.	Belgium
BOS. & HERZ.	Bosnia and Herzegovina
CZECHIA	Czechia (Czech Republic)
KOS.	Kosovo
LIECH.	Liechtenstein
LUX.	Luxembourg
N. MAC.	North Macedonia
MON.	Montenegro
NETH.	Netherlands
NZ	New Zealand
S.M.	San Marino
SLVN.	Slovenia
SWITZ.	Switzerland
U.A.E.	United Arab Emirates
UK	United Kingdom
USA	United States of America
VAT. CITY	Vatican City

WHERE ON EARTH?

OUR WORLD AS YOU'VE NEVER SEEN IT BEFORE

SECOND EDITION
Senior editor Rachel Thompson
Senior art editor Rachael Grady
Senior cartographic editor Simon Mumford
US editor Karyn Gerhard
Designers Chrissy Barnard, Kit Lane
Managing editor Francesca Baines
Managing art editor Philip Letsu
Production editor Gillian Reid
Production controller Samantha Cross
Jacket designer Juthi Seth

FIRST EDITION
Senior editor Rob Houston
Senior art editor Philip Letsu
Senior cartographic editor Simon Mumford
Editors Helen Abramson, Steve Setford, Rona Skene
Designers David Ball, Carol Davis, Mik Gates
Researchers Helen Saunders, Suneha Dutta, Kaiya Shang
Cartography Encompass Graphics, Ed Merritt
Illustrators Adam Benton, Stuart Jackson-Carter
Creative retouching Steve Willis

Picture research Taiyaba Khatoon,
Ashwin Adimari, Martin Copeland
Jacket design Laura Brim, Natasha Rees
Jacket design development manager
Sophia M. Tampakopoulos Turner
Pre-production producer Rebekah Parsons-King
Production controller Mandy Innes
Publisher Andrew Macintyre
Art director Phil Ormerod
Associate publishing director Liz Wheeler
Publishing director Jonathan Metcalf

This American Edition, 2021
First American Edition, 2013
Published in the United States by DK Publishing
1450 Broadway, Suite 801, New York, NY 10018

Copyright © 2013, 2021 Dorling Kindersley Limited
DK, a Division of Penguin Random House LLC
21 22 23 24 25 10 9 8 7 6 5 4 3 2 1
001–323217–Sep/2021

A catalog record for this book
is available from the Library of Congress.
ISBN 978-0-7440-3670-1

DK books are available at special discounts when purchased
in bulk for sales promotions, premiums, fundraising, or educational use.
For details, contact: DK Publishing Special Markets, 1450 Broadway,
Suite 801, New York, NY 10018
SpecialSales@dk.com

Printed and bound in the UAE

For the curious
www.dk.com

MIX
Paper from
responsible sources
FSC™ C018179

This book was made with Forest Stewardship Council™
certified paper—one small step in DK's commitment
to a sustainable future.
For more information go to
www.dk.com/our-green-pledge

CONTENTS

Land, sea, and air

Introduction	6
Earth's crust	8
Earthquakes	10
Mountains	12
Volcanoes	14
Ocean floor	16
Ocean in motion	18
Rivers	20
Craters and meteorites	22
Hot and cold	24
Rain and snow	26
Hurricanes	28
Biomes	30
Forests	32
Deserts	34
Ice	36
Time zones	38

Living world

Introduction	42
Dinosaur fossils	44
Predators	46
Deadly creatures	48
Alien invasion	50
Bird migrations	52
Whales	54
Sharks	56
River monsters	58
Insects	60
World of plants	62
Biodiversity	64
Unique wildlife	66
Endangered animals	68
Extinct animals	70

People and planet

Introduction 74
Where people live 76
Nomads 78
Young and old 80
Health 82
Pandemics 84
Poverty 86
The world's gold 88
Billionaires 90
Food production 92
Food intake 94
Literacy 96
Pollution 98
Garbage and waste 100
Clean water 102
Fossil fuels 104
Alternative energy 106
Climate change 108
Wilderness 110

Engineering and technology

Introduction 114
Air traffic 116
Shipping 118
Railroads 120
Roads 122
Tallest buildings 124
Internet connections 126
Satellites and space junk 128
Armed forces 130

History

Introduction 134
Fossil humans 136
Prehistoric culture 138
Ancient empires 140
Ancient wonders 142
Mummies 144
Medieval wonders 146
Medieval empires 148
Castles 150
Battlegrounds 152
The last empires 154
Revolutions 156
Shipwrecks 158
Industrial wonders 160

Culture

Introduction 164
Languages 166
Holy places 168
Tourism 170
Art 172
Statues 174
Festivals 176
Television 178
Stadiums 180
Motor racing 182
Roller coasters 184
National flags 186

Index 188
Acknowledgments 192

Land, sea, and air

Skeleton Coast, Namibia
The Atlantic Ocean meets the
edge of Africa's Namib Desert
at the Skeleton Coast. Rainfall
here rarely exceeds 0.39 in
(10 mm) per year.

Introduction

Earth is a planet in motion, spinning on its axis as it hurtles through space around the sun. Warmed by the sun's rays, Earth's atmosphere and oceans are always on the move, while heat from the planet's core keeps the hot rock of the interior constantly churning. All of this enables Earth's surface to teem with life.

Churning interior

The rocks in the mantle flow in currents that rise, flow sideways, cool, and then sink. These currents can force the plates of Earth's crust apart or pull sections of the crust back down into the mantle.

Ocean floor splits, while mantle rock rises and creates new crust in the gap

Continent is dragged along by the mantle moving beneath

Mantle moves in slow circles, driven by the core's heat below

Crust is destroyed as it is dragged into the mantle by the sinking current

Water cycle

The sun's heat evaporates sea water, causing it to become water vapor in the air. As it rises and cools, the water vapor condenses into clouds of droplets or ice crystals. As the droplets or crystals grow, they fall as rain or snow. If it falls on land, some runs off the surface to form rivers and lakes, which return water to the oceans. A lot of rain seeps through gaps in the soil and rock. It is called groundwater, and it may stay underground or trickle to the sea. This continuous circulation of water is known as the water cycle.

Earth's structure

If we could take a slice out of Earth, we would see that the planet is made up of layers. At its heart lies a solid inner core, surrounded by a liquid outer core. Both are made mainly of heavy iron. The outer core is enclosed by a deep layer of heavy, very hot, yet solid rock called the mantle. Heat from the core drives currents rising through the mantle that keep the rock moving extremely slowly. The crust—the mantle's cool, hard shell—is made up of a number of rocky plates.

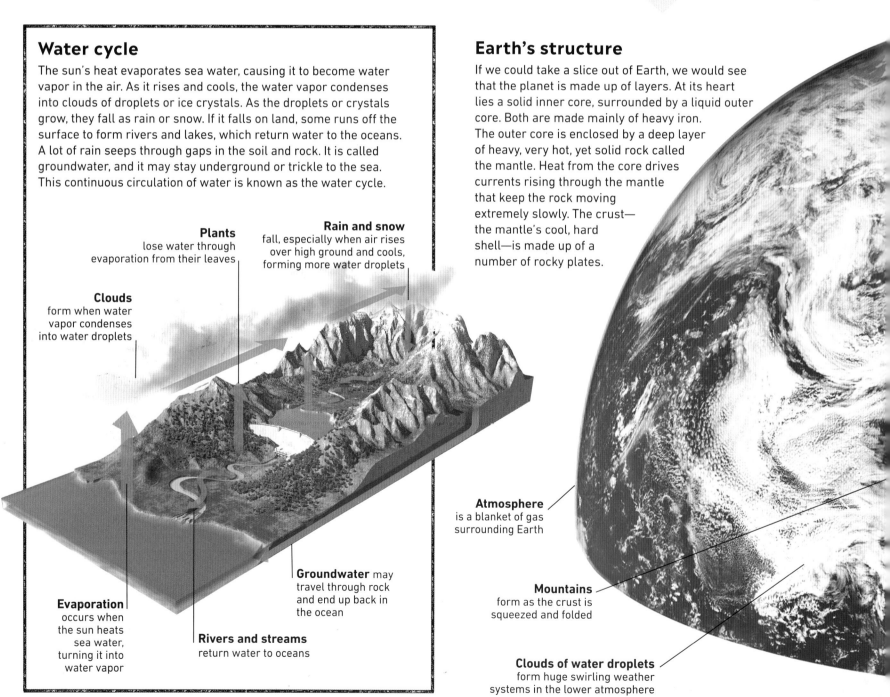

Plants lose water through evaporation from their leaves

Rain and snow fall, especially when air rises over high ground and cools, forming more water droplets

Clouds form when water vapor condenses into water droplets

Evaporation occurs when the sun heats sea water, turning it into water vapor

Rivers and streams return water to oceans

Groundwater may travel through rock and end up back in the ocean

Atmosphere is a blanket of gas surrounding Earth

Mountains form as the crust is squeezed and folded

Clouds of water droplets form huge swirling weather systems in the lower atmosphere

AN AVERAGE WATER MOLECULE SPENDS ABOUT 3,200 YEARS IN THE

The sun's energy

In the tropics, near the equator, the sun's rays strike Earth at a steep angle, so the energy is very concentrated. But near the poles, sunlight hits the surface at a narrow angle. This spreads the sun's energy, giving a weak heating effect. The result is that polar regions are much colder than tropical zones, allowing ice to form in the Arctic and Antarctic. The difference in the solar heating at different latitudes sets bodies of air and seawater in motion, driving winds and ocean currents.

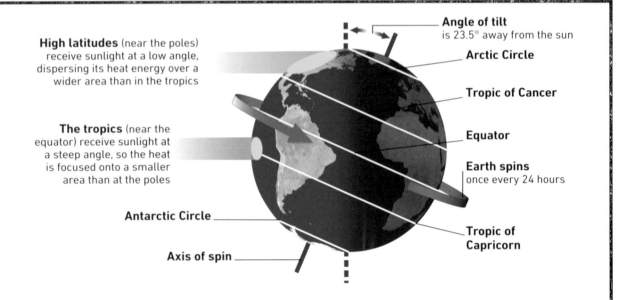

High latitudes (near the poles) receive sunlight at a low angle, dispersing its heat energy over a wider area than in the tropics

The tropics (near the equator) receive sunlight at a steep angle, so the heat is focused onto a smaller area than at the poles

Antarctic Circle

Axis of spin

Angle of tilt is 23.5° away from the sun

Arctic Circle

Tropic of Cancer

Equator

Earth spins once every 24 hours

Tropic of Capricorn

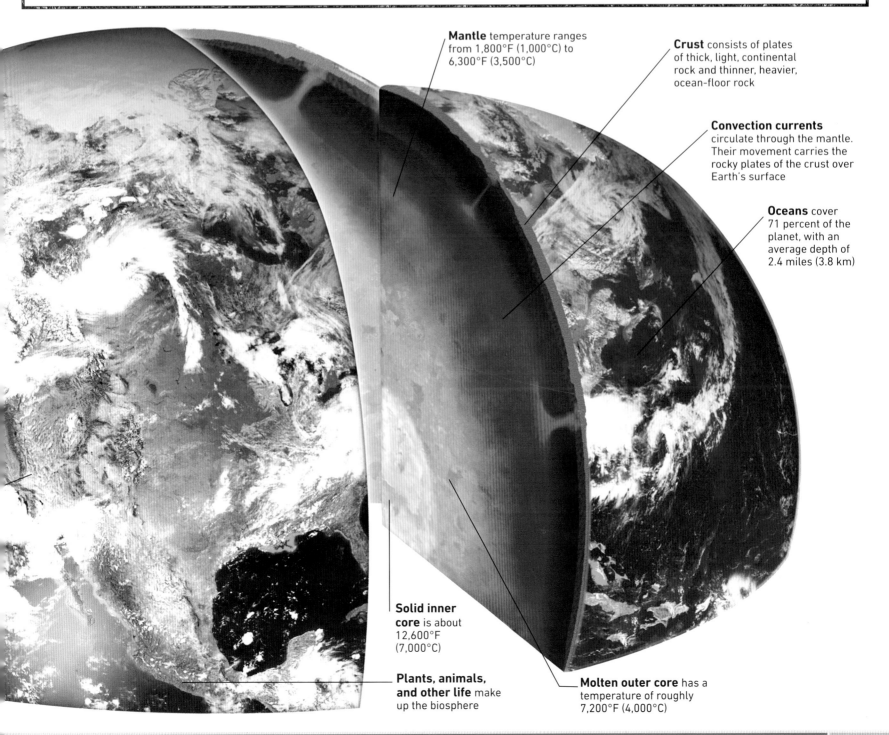

Mantle temperature ranges from 1,800°F (1,000°C) to 6,300°F (3,500°C)

Crust consists of plates of thick, light, continental rock and thinner, heavier, ocean-floor rock

Convection currents circulate through the mantle. Their movement carries the rocky plates of the crust over Earth's surface

Oceans cover 71 percent of the planet, with an average depth of 2.4 miles (3.8 km)

Solid inner core is about 12,600°F (7,000°C)

Plants, animals, and other life make up the biosphere

Molten outer core has a temperature of roughly 7,200°F (4,000°C)

OCEAN BEFORE EVAPORATION RELEASES IT INTO THE ATMOSPHERE.

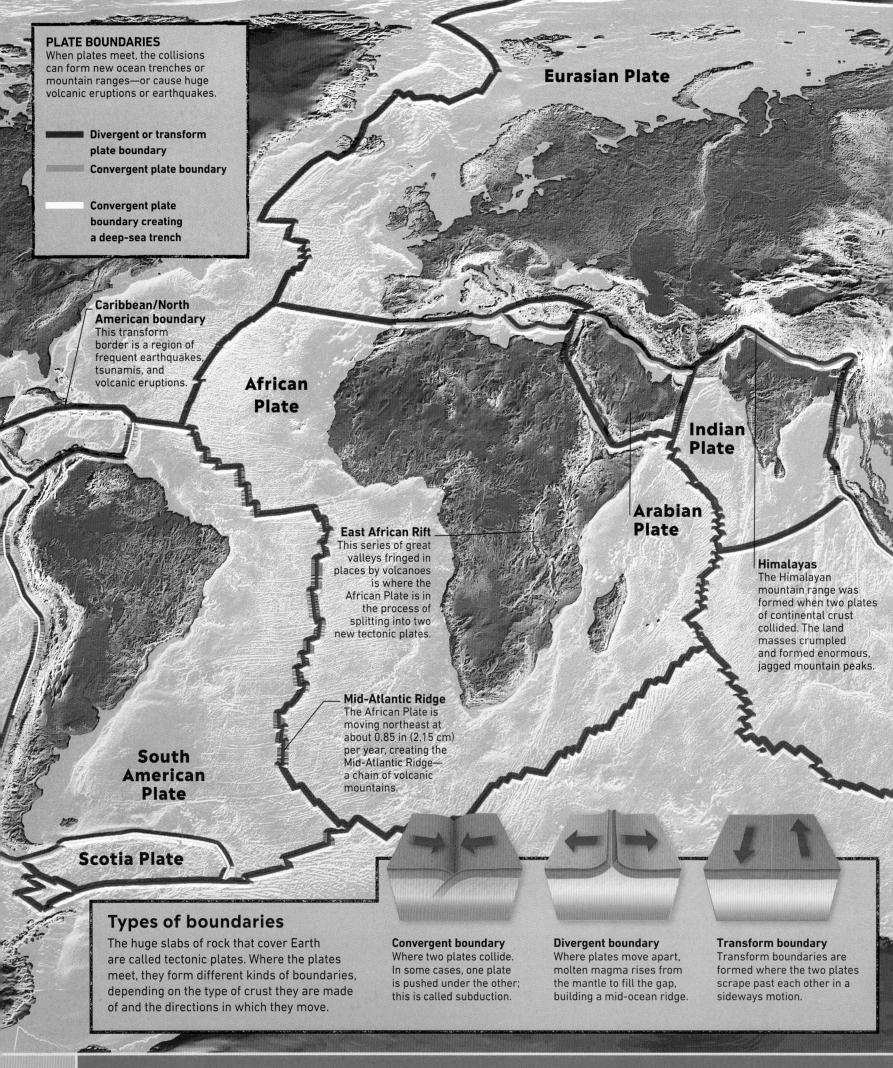

PLATE BOUNDARIES
When plates meet, the collisions can form new ocean trenches or mountain ranges—or cause huge volcanic eruptions or earthquakes.

━━━ Divergent or transform plate boundary

━━━ Convergent plate boundary

━━━ Convergent plate boundary creating a deep-sea trench

Eurasian Plate

Caribbean/North American boundary
This transform border is a region of frequent earthquakes, tsunamis, and volcanic eruptions.

African Plate

Indian Plate

Arabian Plate

East African Rift
This series of great valleys fringed in places by volcanoes is where the African Plate is in the process of splitting into two new tectonic plates.

Himalayas
The Himalayan mountain range was formed when two plates of continental crust collided. The land masses crumpled and formed enormous, jagged mountain peaks.

Mid-Atlantic Ridge
The African Plate is moving northeast at about 0.85 in (2.15 cm) per year, creating the Mid-Atlantic Ridge—a chain of volcanic mountains.

South American Plate

Scotia Plate

Types of boundaries

The huge slabs of rock that cover Earth are called tectonic plates. Where the plates meet, they form different kinds of boundaries, depending on the type of crust they are made of and the directions in which they move.

Convergent boundary
Where two plates collide. In some cases, one plate is pushed under the other; this is called subduction.

Divergent boundary
Where plates move apart, molten magma rises from the mantle to fill the gap, building a mid-ocean ridge.

Transform boundary
Transform boundaries are formed where the two plates scrape past each other in a sideways motion.

RELATIVE TO ITS SIZE, THE EARTH'S CRUST IS THINNER THAN THE

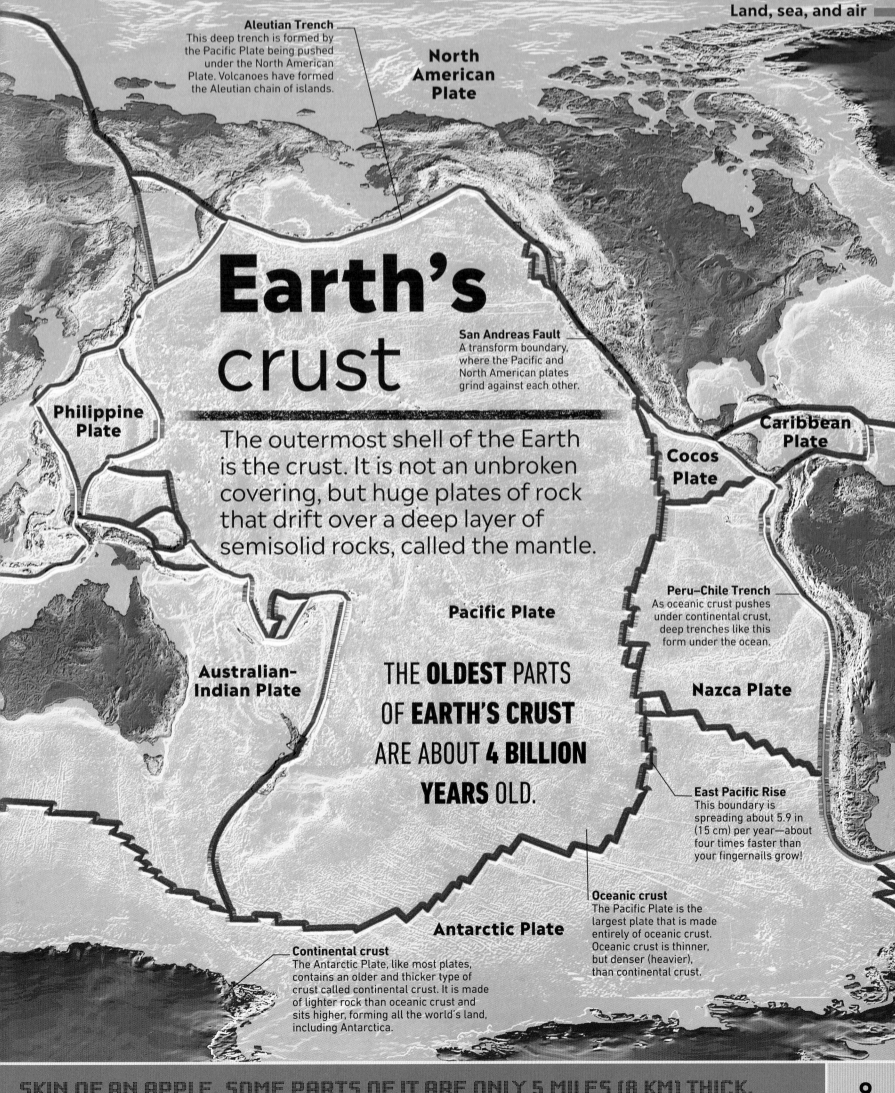

Aleutian Trench
This deep trench is formed by the Pacific Plate being pushed under the North American Plate. Volcanoes have formed the Aleutian chain of islands.

North American Plate

Earth's crust

San Andreas Fault
A transform boundary, where the Pacific and North American plates grind against each other.

Philippine Plate

Caribbean Plate

Cocos Plate

The outermost shell of the Earth is the crust. It is not an unbroken covering, but huge plates of rock that drift over a deep layer of semisolid rocks, called the mantle.

Peru–Chile Trench
As oceanic crust pushes under continental crust, deep trenches like this form under the ocean.

Pacific Plate

Australian-Indian Plate

THE **OLDEST** PARTS OF **EARTH'S CRUST** ARE ABOUT **4 BILLION YEARS** OLD.

Nazca Plate

East Pacific Rise
This boundary is spreading about 5.9 in (15 cm) per year—about four times faster than your fingernails grow!

Oceanic crust
The Pacific Plate is the largest plate that is made entirely of oceanic crust. Oceanic crust is thinner, but denser (heavier), than continental crust.

Antarctic Plate

Continental crust
The Antarctic Plate, like most plates, contains an older and thicker type of crust called continental crust. It is made of lighter rock than oceanic crust and sits higher, forming all the world's land, including Antarctica.

Strongest earthquakes since 1900

(1) Valdivia, Chile—May 22, 1960
This earthquake measured 9.5 in magnitude. It killed 1,655 people and caused a tsunami that hit Japan, the Philippines, and the US.

(2) Prince William Sound, Alaska—1964
This 9.2-magnitude earthquake hit Alaska on March 27. While it killed 15 people, it caused a tsunami that killed another 113.

(3) Indian Ocean—December 26, 2004
Occurring at sea, this 9.1-magnitude earthquake caused a tsunami that killed 227,898 people and affected 1.7 million more.

(4) Kamchatka, Russia—November 4, 1952
This 9.0-magnitude earthquake sent a tsunami across the Pacific. In Hawaii, no human lives were lost, but six cows died.

(5) Tohoku, Japan—March 11, 2011
This 9.0-magnitude earthquake and tsunami killed more than 15,000 people and destroyed a nuclear power plant.

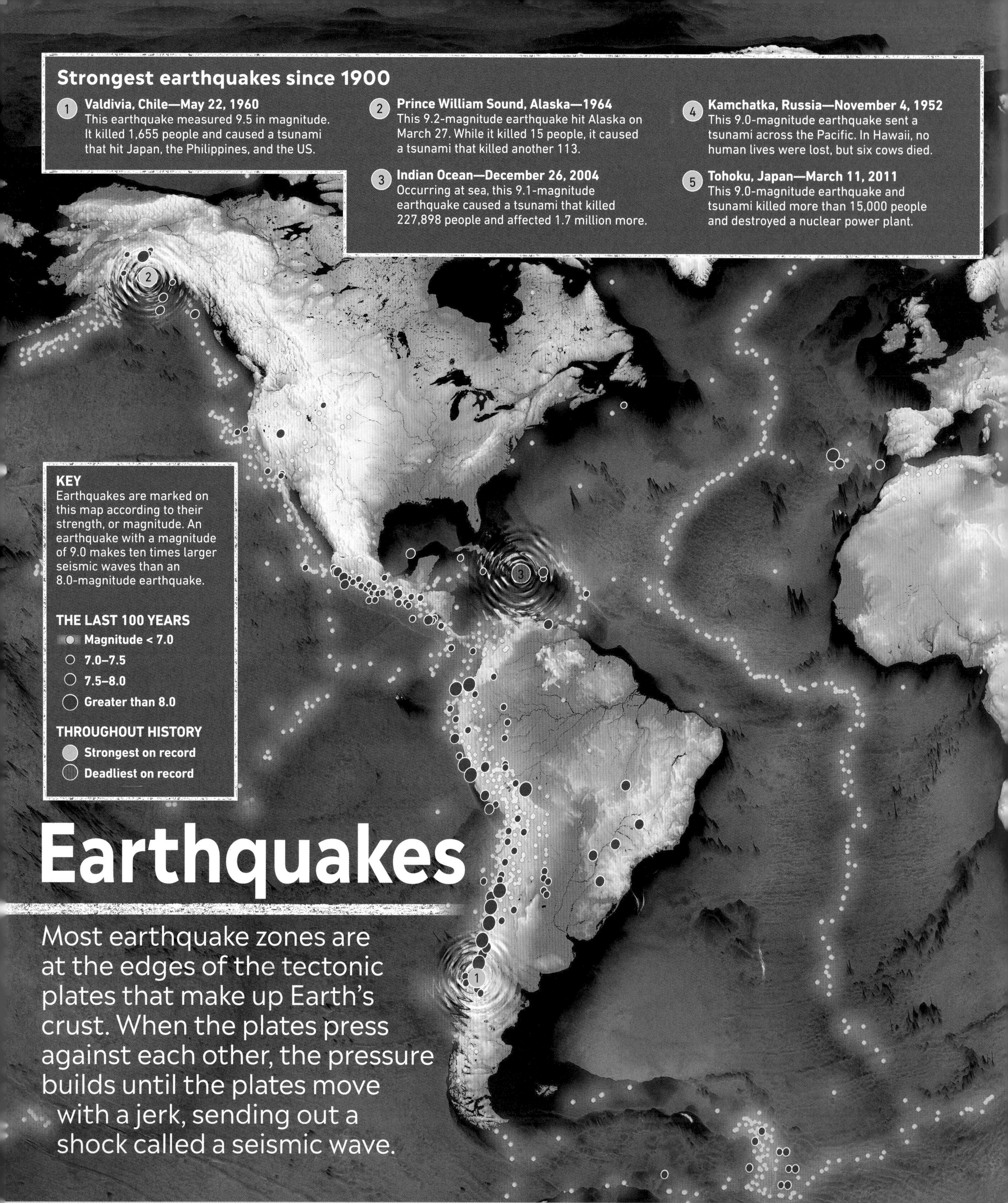

KEY

Earthquakes are marked on this map according to their strength, or magnitude. An earthquake with a magnitude of 9.0 makes ten times larger seismic waves than an 8.0-magnitude earthquake.

THE LAST 100 YEARS
- Magnitude < 7.0
- 7.0–7.5
- 7.5–8.0
- Greater than 8.0

THROUGHOUT HISTORY
- Strongest on record
- Deadliest on record

Earthquakes

Most earthquake zones are at the edges of the tectonic plates that make up Earth's crust. When the plates press against each other, the pressure builds until the plates move with a jerk, sending out a shock called a seismic wave.

THE ASTEROID IMPACT THAT WIPED OUT THE DINOSAURS 65 MILLION YEARS

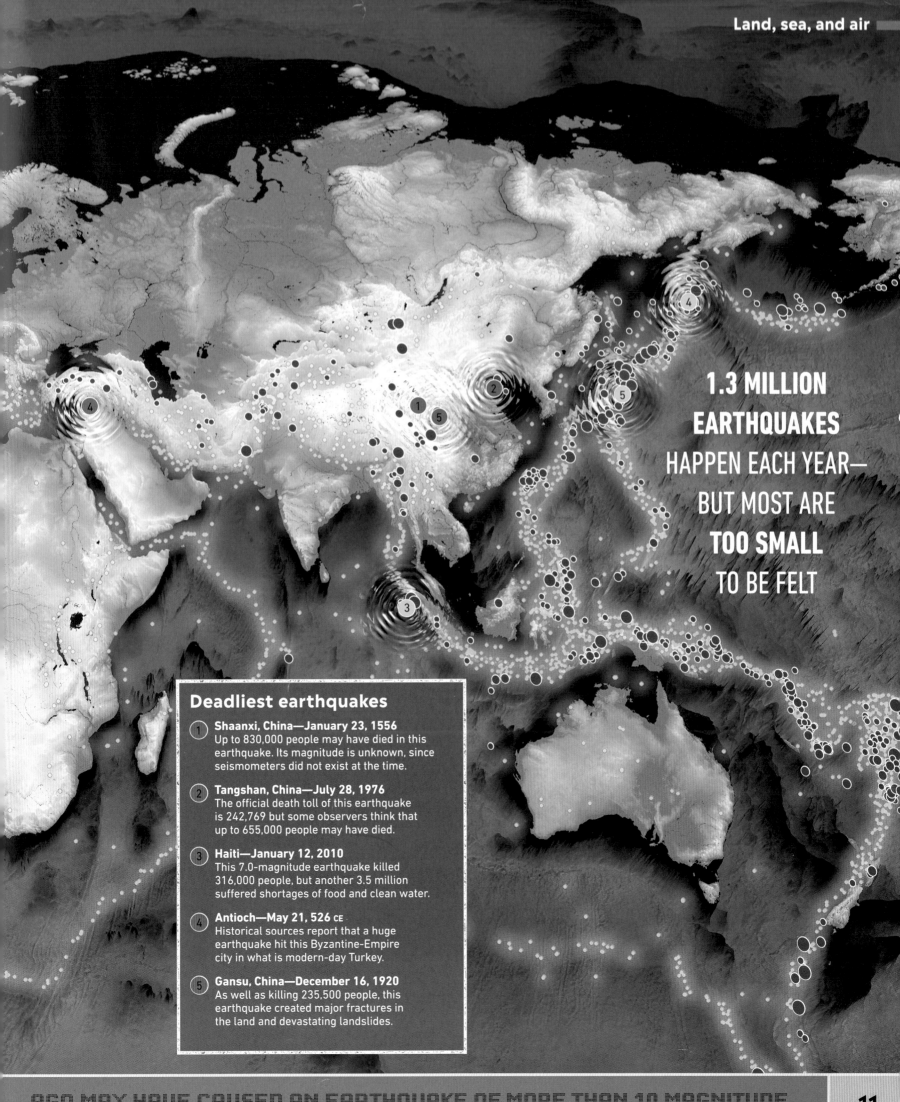

1.3 MILLION EARTHQUAKES HAPPEN EACH YEAR— BUT MOST ARE **TOO SMALL** TO BE FELT

Deadliest earthquakes

1. **Shaanxi, China—January 23, 1556**
   Up to 830,000 people may have died in this earthquake. Its magnitude is unknown, since seismometers did not exist at the time.

2. **Tangshan, China—July 28, 1976**
   The official death toll of this earthquake is 242,769 but some observers think that up to 655,000 people may have died.

3. **Haiti—January 12, 2010**
   This 7.0-magnitude earthquake killed 316,000 people, but another 3.5 million suffered shortages of food and clean water.

4. **Antioch—May 21, 526 CE**
   Historical sources report that a huge earthquake hit this Byzantine-Empire city in what is modern-day Turkey.

5. **Gansu, China—December 16, 1920**
   As well as killing 235,500 people, this earthquake created major fractures in the land and devastating landslides.

AGO MAY HAVE CAUSED AN EARTHQUAKE OF MORE THAN 10 MAGNITUDE.

11

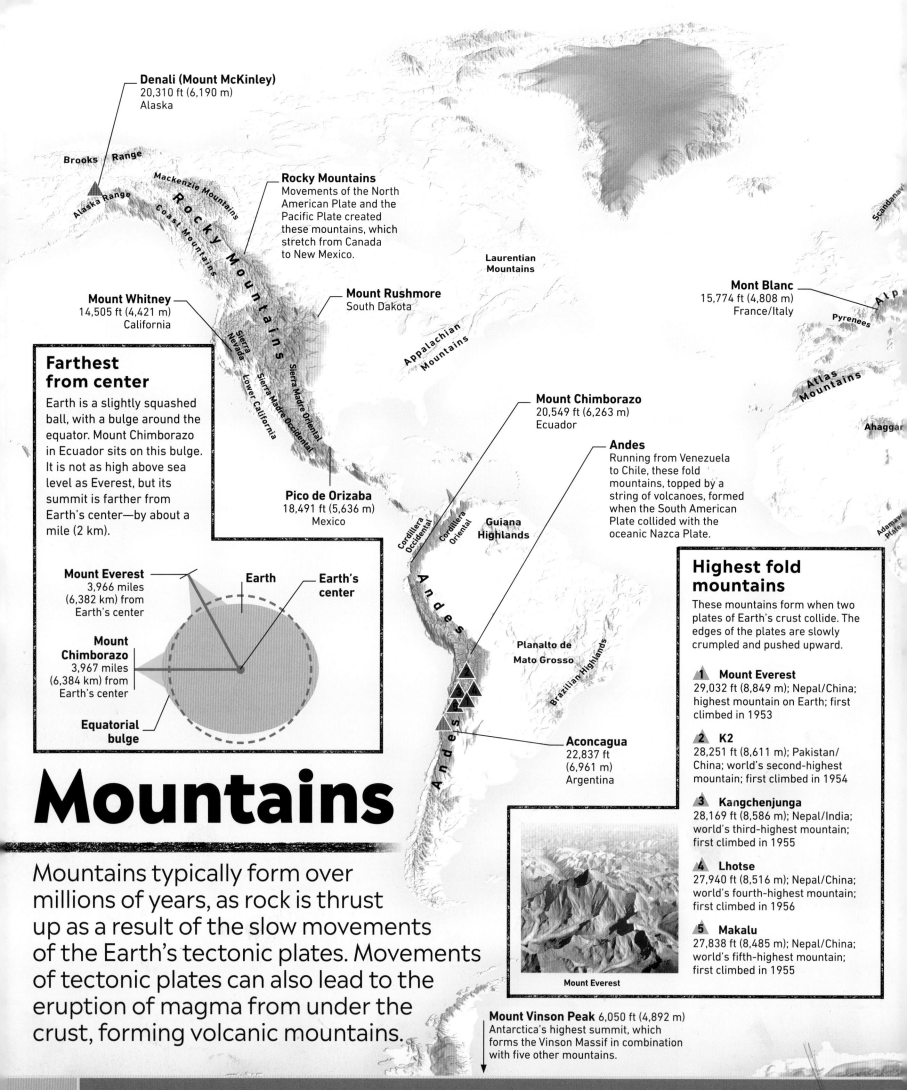

Denali (Mount McKinley)
20,310 ft (6,190 m)
Alaska

Brooks Range

Mackenzie Mountains

Coast Mountains

Rocky Mountains

Alaska Range

Rocky Mountains
Movements of the North American Plate and the Pacific Plate created these mountains, which stretch from Canada to New Mexico.

Laurentian Mountains

Mont Blanc
15,774 ft (4,808 m)
France/Italy

Alps

Pyrenees

Scandanavia

Mount Whitney
14,505 ft (4,421 m)
California

Sierra Nevada

Sierra Madre Occidental

Lower California

Sierra Madre Oriental

Mount Rushmore
South Dakota

Appalachian Mountains

Atlas Mountains

Ahaggar

Farthest from center

Earth is a slightly squashed ball, with a bulge around the equator. Mount Chimborazo in Ecuador sits on this bulge. It is not as high above sea level as Everest, but its summit is farther from Earth's center—by about a mile (2 km).

Mount Chimborazo
20,549 ft (6,263 m)
Ecuador

Andes
Running from Venezuela to Chile, these fold mountains, topped by a string of volcanoes, formed when the South American Plate collided with the oceanic Nazca Plate.

Pico de Orizaba
18,491 ft (5,636 m)
Mexico

Cordillera Occidental

Cordillera Oriental

Guiana Highlands

Adamawa Plateau

Mount Everest
3,966 miles (6,382 km) from Earth's center

Earth

Earth's center

Mount Chimborazo
3,967 miles (6,384 km) from Earth's center

Equatorial bulge

Andes

Planalto de Mato Grosso

Brazilian Highlands

Highest fold mountains

These mountains form when two plates of Earth's crust collide. The edges of the plates are slowly crumpled and pushed upward.

1 **Mount Everest**
29,032 ft (8,849 m); Nepal/China; highest mountain on Earth; first climbed in 1953

2 **K2**
28,251 ft (8,611 m); Pakistan/China; world's second-highest mountain; first climbed in 1954

3 **Kangchenjunga**
28,169 ft (8,586 m); Nepal/India; world's third-highest mountain; first climbed in 1955

4 **Lhotse**
27,940 ft (8,516 m); Nepal/China; world's fourth-highest mountain; first climbed in 1956

5 **Makalu**
27,838 ft (8,485 m); Nepal/China; world's fifth-highest mountain; first climbed in 1955

Aconcagua
22,837 ft (6,961 m)
Argentina

Mountains

Mountains typically form over millions of years, as rock is thrust up as a result of the slow movements of the Earth's tectonic plates. Movements of tectonic plates can also lead to the eruption of magma from under the crust, forming volcanic mountains.

Mount Everest

Mount Vinson Peak 6,050 ft (4,892 m)
Antarctica's highest summit, which forms the Vinson Massif in combination with five other mountains.

THE TALLEST KNOWN MOUNTAIN ANYWHERE IS OLYMPUS MONS,

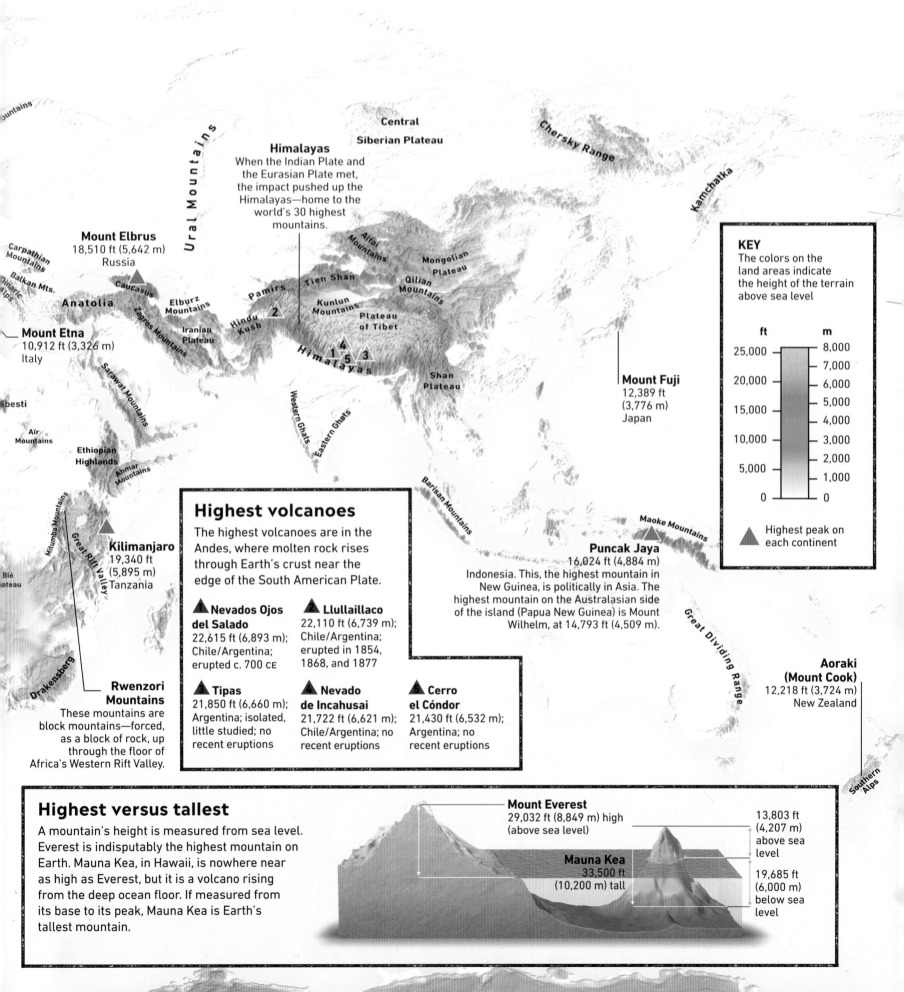

Himalayas
When the Indian Plate and the Eurasian Plate met, the impact pushed up the Himalayas—home to the world's 30 highest mountains.

Central Siberian Plateau

Chersky Range

Ural Mountains

Mount Elbrus
18,510 ft (5,642 m)
Russia

Carpathian Mountains

Balkan Mts.

Dinaric Alps

Anatolia

Caucasus

Elburz Mountains

Zagros Mountains

Iranian Plateau

Hindu Kush

Pamirs

Tien Shan

Kunlun Mountains

Altai Mountains

Mongolian Plateau

Qilian Mountains

Plateau of Tibet

Himalayas

Shan Plateau

Kamchatka

Mount Etna
10,912 ft (3,326 m)
Italy

Sarawat Mountains

Tibesti

Aïr Mountains

Ethiopian Highlands

Ahmar Mountains

Western Ghats

Eastern Ghats

Mount Fuji
12,389 ft
(3,776 m)
Japan

Mitumba Mountains

Great Rift Valley

Kilimanjaro
19,340 ft
(5,895 m)
Tanzania

Bié Plateau

Drakensberg

Barisan Mountains

Maoke Mountains

Puncak Jaya
16,024 ft (4,884 m)
Indonesia. This, the highest mountain in New Guinea, is politically in Asia. The highest mountain on the Australasian side of the island (Papua New Guinea) is Mount Wilhelm, at 14,793 ft (4,509 m).

Great Dividing Range

**Aoraki
(Mount Cook)**
12,218 ft (3,724 m)
New Zealand

Southern Alps

Rwenzori Mountains
These mountains are block mountains—forced, as a block of rock, up through the floor of Africa's Western Rift Valley.

Highest volcanoes

The highest volcanoes are in the Andes, where molten rock rises through Earth's crust near the edge of the South American Plate.

▲ Nevados Ojos del Salado
22,615 ft (6,893 m);
Chile/Argentina;
erupted c. 700 CE

▲ Llullaillaco
22,110 ft (6,739 m);
Chile/Argentina;
erupted in 1854,
1868, and 1877

▲ Tipas
21,850 ft (6,660 m);
Argentina; isolated,
little studied; no
recent eruptions

▲ Nevado de Incahuasi
21,722 ft (6,621 m);
Chile/Argentina; no
recent eruptions

▲ Cerro el Cóndor
21,430 ft (6,532 m);
Argentina; no
recent eruptions

KEY
The colors on the land areas indicate the height of the terrain above sea level

ft	m
25,000	8,000
	7,000
	6,000
20,000	5,000
15,000	4,000
	3,000
10,000	2,000
	1,000
5,000	
0	0

▲ Highest peak on each continent

Highest versus tallest

A mountain's height is measured from sea level. Everest is indisputably the highest mountain on Earth. Mauna Kea, in Hawaii, is nowhere near as high as Everest, but it is a volcano rising from the deep ocean floor. If measured from its base to its peak, Mauna Kea is Earth's tallest mountain.

Mount Everest
29,032 ft (8,849 m) high
(above sea level)

Mauna Kea
33,500 ft
(10,200 m) tall

13,803 ft
(4,207 m)
above sea
level

19,685 ft
(6,000 m)
below sea
level

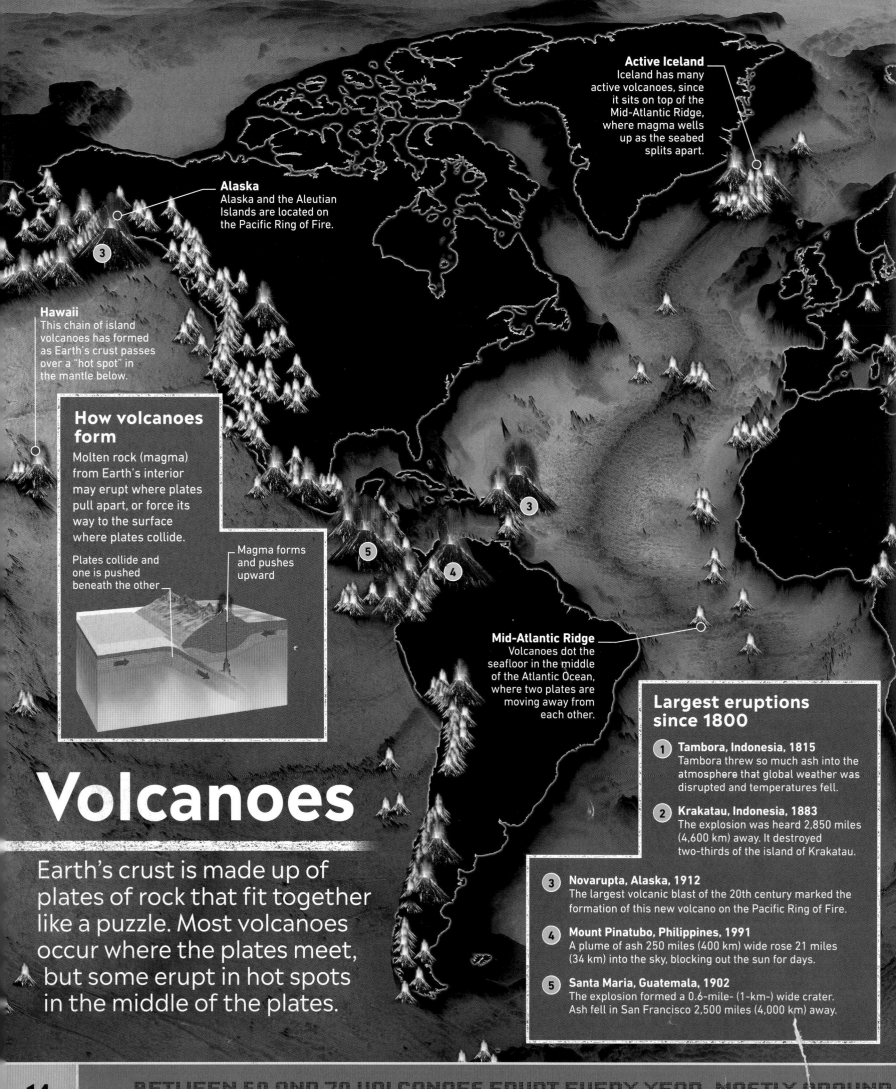

Volcanoes

Active Iceland
Iceland has many active volcanoes, since it sits on top of the Mid-Atlantic Ridge, where magma wells up as the seabed splits apart.

Alaska
Alaska and the Aleutian Islands are located on the Pacific Ring of Fire.

Hawaii
This chain of island volcanoes has formed as Earth's crust passes over a "hot spot" in the mantle below.

How volcanoes form

Molten rock (magma) from Earth's interior may erupt where plates pull apart, or force its way to the surface where plates collide.

Plates collide and one is pushed beneath the other

Magma forms and pushes upward

Mid-Atlantic Ridge
Volcanoes dot the seafloor in the middle of the Atlantic Ocean, where two plates are moving away from each other.

Earth's crust is made up of plates of rock that fit together like a puzzle. Most volcanoes occur where the plates meet, but some erupt in hot spots in the middle of the plates.

Largest eruptions since 1800

1. **Tambora, Indonesia, 1815**
Tambora threw so much ash into the atmosphere that global weather was disrupted and temperatures fell.

2. **Krakatau, Indonesia, 1883**
The explosion was heard 2,850 miles (4,600 km) away. It destroyed two-thirds of the island of Krakatau.

3. **Novarupta, Alaska, 1912**
The largest volcanic blast of the 20th century marked the formation of this new volcano on the Pacific Ring of Fire.

4. **Mount Pinatubo, Philippines, 1991**
A plume of ash 250 miles (400 km) wide rose 21 miles (34 km) into the sky, blocking out the sun for days.

5. **Santa Maria, Guatemala, 1902**
The explosion formed a 0.6-mile- (1-km-) wide crater. Ash fell in San Francisco 2,500 miles (4,000 km) away.

BETWEEN 50 AND 70 VOLCANOES ERUPT EVERY YEAR, MOSTLY AROUND

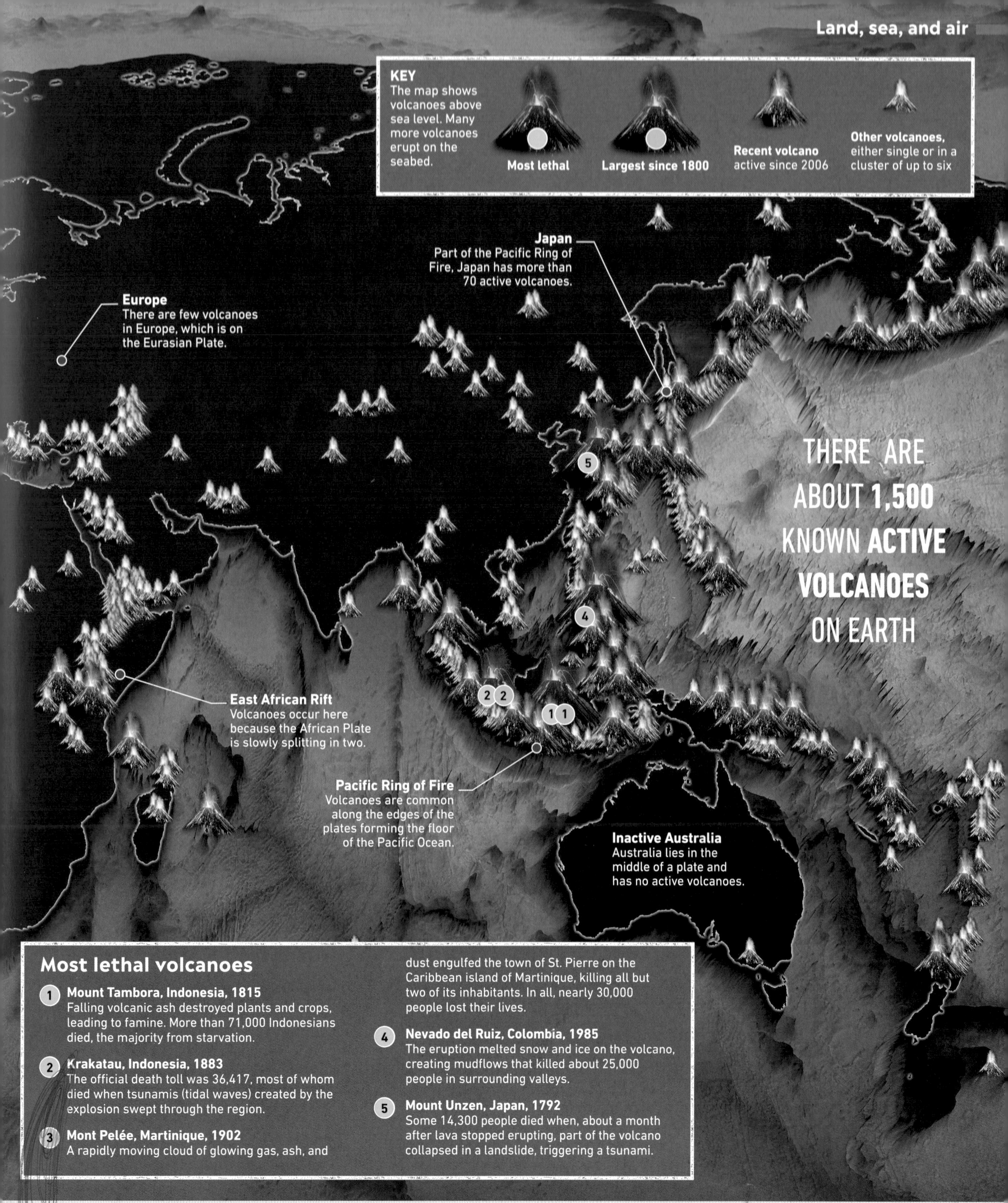

KEY
The map shows volcanoes above sea level. Many more volcanoes erupt on the seabed.

Most lethal

Largest since 1800

Recent volcano active since 2006

Other volcanoes, either single or in a cluster of up to six

Japan
Part of the Pacific Ring of Fire, Japan has more than 70 active volcanoes.

Europe
There are few volcanoes in Europe, which is on the Eurasian Plate.

THERE ARE ABOUT **1,500** KNOWN **ACTIVE VOLCANOES** ON EARTH

East African Rift
Volcanoes occur here because the African Plate is slowly splitting in two.

Pacific Ring of Fire
Volcanoes are common along the edges of the plates forming the floor of the Pacific Ocean.

Inactive Australia
Australia lies in the middle of a plate and has no active volcanoes.

Most lethal volcanoes

1 Mount Tambora, Indonesia, 1815
Falling volcanic ash destroyed plants and crops, leading to famine. More than 71,000 Indonesians died, the majority from starvation.

2 Krakatau, Indonesia, 1883
The official death toll was 36,417, most of whom died when tsunamis (tidal waves) created by the explosion swept through the region.

3 Mont Pelée, Martinique, 1902
A rapidly moving cloud of glowing gas, ash, and

dust engulfed the town of St. Pierre on the Caribbean island of Martinique, killing all but two of its inhabitants. In all, nearly 30,000 people lost their lives.

4 Nevado del Ruiz, Colombia, 1985
The eruption melted snow and ice on the volcano, creating mudflows that killed about 25,000 people in surrounding valleys.

5 Mount Unzen, Japan, 1792
Some 14,300 people died when, about a month after lava stopped erupting, part of the volcano collapsed in a landslide, triggering a tsunami.

THE EDGE OF THE PACIFIC. ABOUT 20 VOLCANOES ARE ERUPTING NOW.

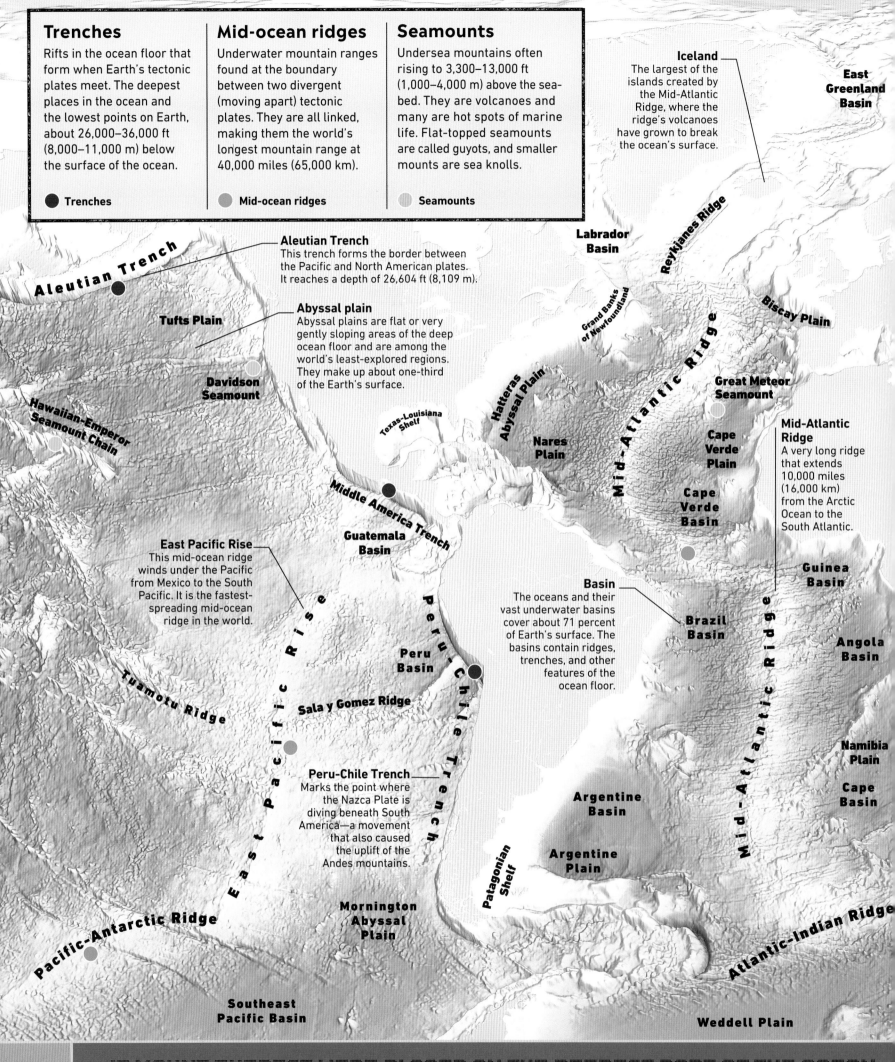

Trenches

Rifts in the ocean floor that form when Earth's tectonic plates meet. The deepest places in the ocean and the lowest points on Earth, about 26,000–36,000 ft (8,000–11,000 m) below the surface of the ocean.

● Trenches

Mid-ocean ridges

Underwater mountain ranges found at the boundary between two divergent (moving apart) tectonic plates. They are all linked, making them the world's longest mountain range at 40,000 miles (65,000 km).

● Mid-ocean ridges

Seamounts

Undersea mountains often rising to 3,300–13,000 ft (1,000–4,000 m) above the sea-bed. They are volcanoes and many are hot spots of marine life. Flat-topped seamounts are called guyots, and smaller mounts are sea knolls.

● Seamounts

Iceland
The largest of the islands created by the Mid-Atlantic Ridge, where the ridge's volcanoes have grown to break the ocean's surface.

East Greenland Basin

Aleutian Trench
This trench forms the border between the Pacific and North American plates. It reaches a depth of 26,604 ft (8,109 m).

Abyssal plain
Abyssal plains are flat or very gently sloping areas of the deep ocean floor and are among the world's least-explored regions. They make up about one-third of the Earth's surface.

East Pacific Rise
This mid-ocean ridge winds under the Pacific from Mexico to the South Pacific. It is the fastest-spreading mid-ocean ridge in the world.

Basin
The oceans and their vast underwater basins cover about 71 percent of Earth's surface. The basins contain ridges, trenches, and other features of the ocean floor.

Peru-Chile Trench
Marks the point where the Nazca Plate is diving beneath South America—a movement that also caused the uplift of the Andes mountains.

Aleutian Trench

Tufts Plain

Davidson Seamount

Hawaiian-Emperor Seamount Chain

Middle America Trench

Guatemala Basin

Texas-Louisiana Shelf

Nares Plain

Hatteras Abyssal Plain

Labrador Basin

Reykjanes Ridge

Grand Banks of Newfoundland

Biscay Plain

Mid-Atlantic Ridge

Great Meteor Seamount

Cape Verde Plain

Cape Verde Basin

Mid-Atlantic Ridge
A very long ridge that extends 10,000 miles (16,000 km) from the Arctic Ocean to the South Atlantic.

Guinea Basin

Tuamotu Ridge

East Pacific Rise

Peru-Chile Trench

Peru Basin

Sala y Gomez Ridge

Brazil Basin

Angola Basin

Namibia Plain

Cape Basin

Argentine Basin

Argentine Plain

Patagonian Shelf

Mornington Abyssal Plain

Pacific-Antarctic Ridge

Mid-Atlantic Ridge

Atlantic-Indian Ridge

Southeast Pacific Basin

Weddell Plain

IF MOUNT EVEREST WERE PLACED ON THE DEEPEST PART OF THE OCEAN

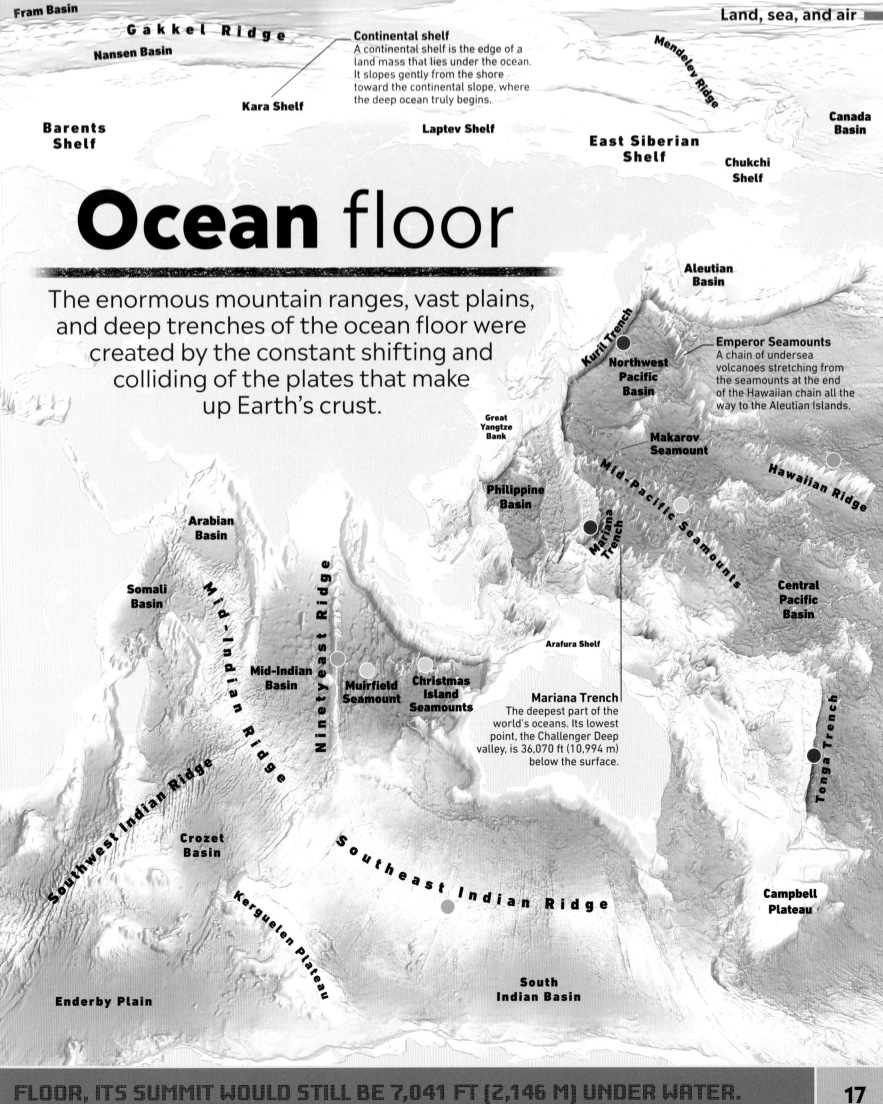

Fram Basin

G a k k e l   R i d g e

Nansen Basin

Mendelev Ridge

Continental shelf
A continental shelf is the edge of a
land mass that lies under the ocean.
It slopes gently from the shore
toward the continental slope, where
the deep ocean truly begins.

Kara Shelf

Canada
Basin

Barents
Shelf

Laptev Shelf

East Siberian
Shelf

Chukchi
Shelf

Ocean floor

Aleutian
Basin

The enormous mountain ranges, vast plains,
and deep trenches of the ocean floor were
created by the constant shifting and
colliding of the plates that make
up Earth's crust.

Kuril Trench

Northwest
Pacific
Basin

Emperor Seamounts
A chain of undersea
volcanoes stretching from
the seamounts at the end
of the Hawaiian chain all the
way to the Aleutian Islands.

Makarov
Seamount

Hawaiian Ridge

Great
Yangtze
Bank

Mid-Pacific Seamounts

Arabian
Basin

Philippine
Basin

Mariana
Trench

Central
Pacific
Basin

Somali
Basin

Mid-Indian Ridge

Mid-Indian
Basin

Ninetyeast Ridge

Muirfield
Seamount

Christmas
Island
Seamounts

Arafura Shelf

Mariana Trench
The deepest part of the
world's oceans. Its lowest
point, the Challenger Deep
valley, is 36,070 ft (10,994 m)
below the surface.

Tonga Trench

Southwest Indian Ridge

Crozet
Basin

Southeast Indian Ridge

Kerguelen Plateau

Campbell
Plateau

Enderby Plain

South
Indian Basin

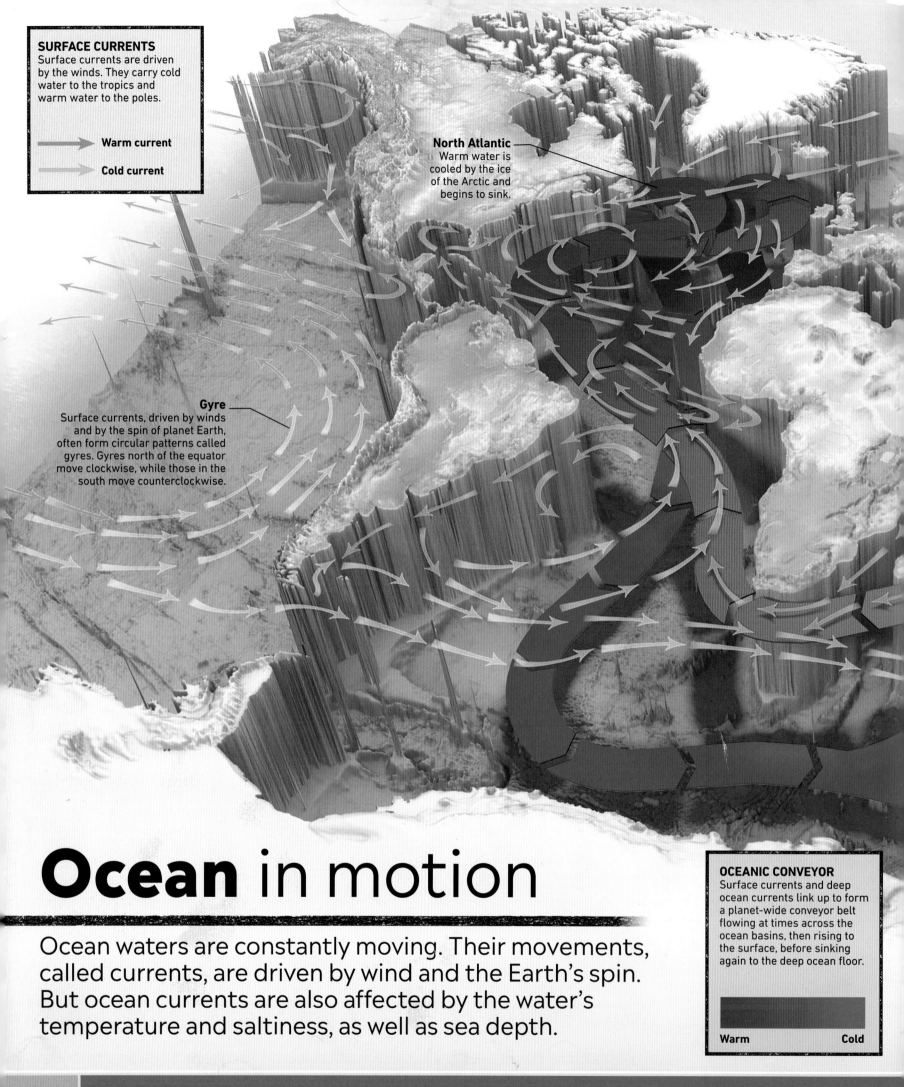

SURFACE CURRENTS
Surface currents are driven by the winds. They carry cold water to the tropics and warm water to the poles.

➤ **Warm current**

➤ **Cold current**

North Atlantic
Warm water is cooled by the ice of the Arctic and begins to sink.

Gyre
Surface currents, driven by winds and by the spin of planet Earth, often form circular patterns called gyres. Gyres north of the equator move clockwise, while those in the south move counterclockwise.

Ocean in motion

Ocean waters are constantly moving. Their movements, called currents, are driven by wind and the Earth's spin. But ocean currents are also affected by the water's temperature and saltiness, as well as sea depth.

OCEANIC CONVEYOR
Surface currents and deep ocean currents link up to form a planet-wide conveyor belt flowing at times across the ocean basins, then rising to the surface, before sinking again to the deep ocean floor.

Warm   Cold

   IT CAN TAKE UP TO 1,000 YEARS FOR OCEAN CURRENTS TO

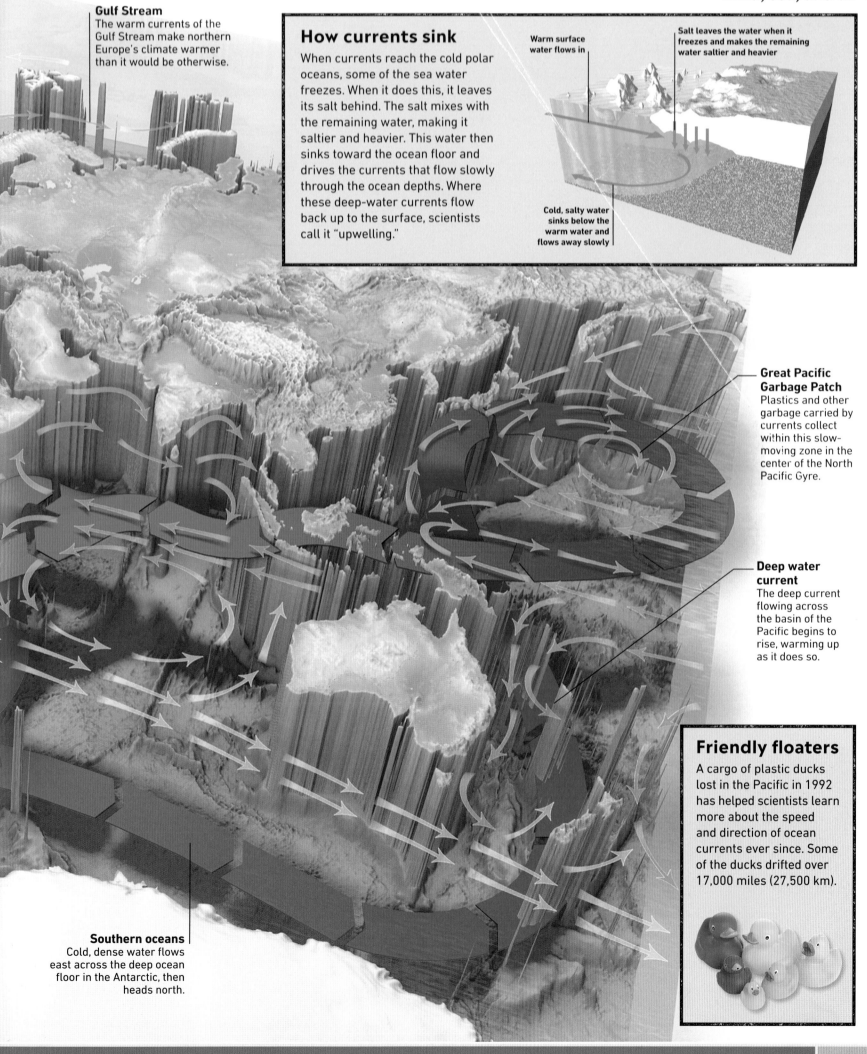

Gulf Stream
The warm currents of the Gulf Stream make northern Europe's climate warmer than it would be otherwise.

How currents sink

When currents reach the cold polar oceans, some of the sea water freezes. When it does this, it leaves its salt behind. The salt mixes with the remaining water, making it saltier and heavier. This water then sinks toward the ocean floor and drives the currents that flow slowly through the ocean depths. Where these deep-water currents flow back up to the surface, scientists call it "upwelling."

Warm surface water flows in

Salt leaves the water when it freezes and makes the remaining water saltier and heavier

Cold, salty water sinks below the warm water and flows away slowly

Great Pacific Garbage Patch
Plastics and other garbage carried by currents collect within this slow-moving zone in the center of the North Pacific Gyre.

Deep water current
The deep current flowing across the basin of the Pacific begins to rise, warming up as it does so.

Friendly floaters

A cargo of plastic ducks lost in the Pacific in 1992 has helped scientists learn more about the speed and direction of ocean currents ever since. Some of the ducks drifted over 17,000 miles (27,500 km).

Southern oceans
Cold, dense water flows east across the deep ocean floor in the Antarctic, then heads north.

Rivers

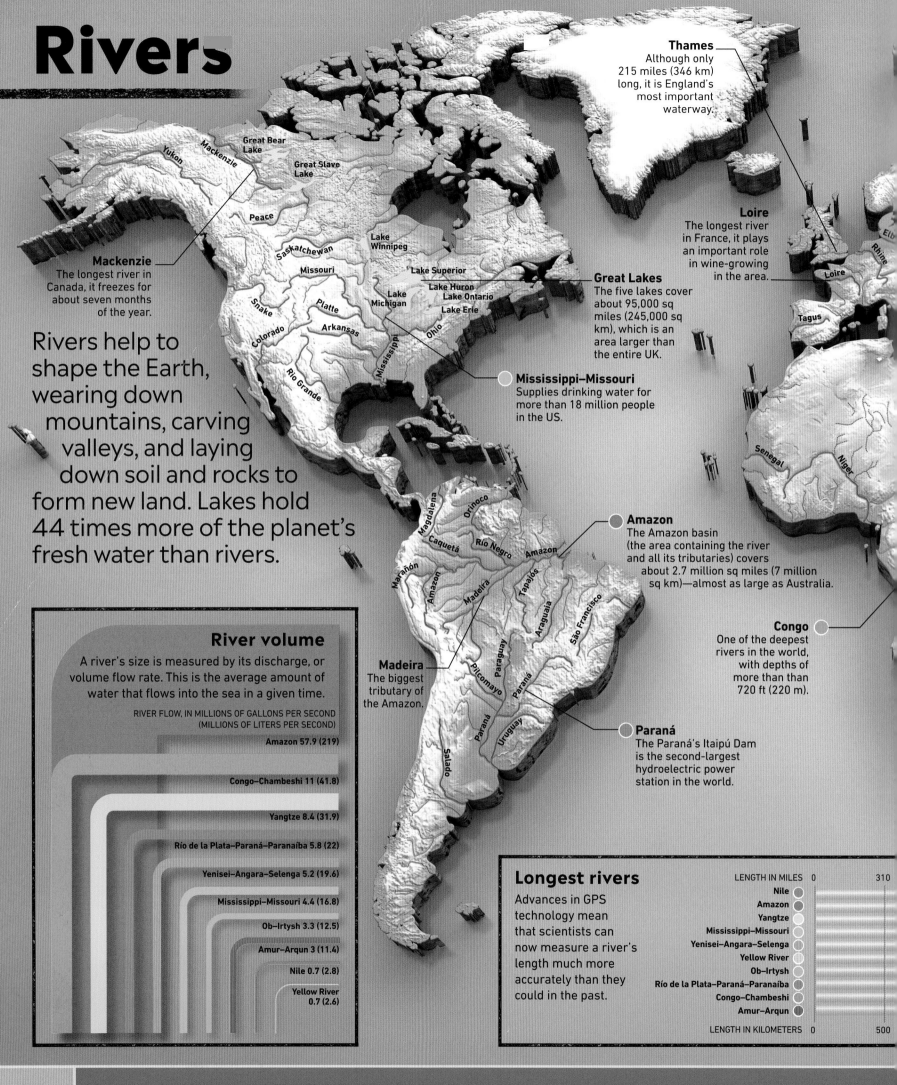

Rivers help to shape the Earth, wearing down mountains, carving valleys, and laying down soil and rocks to form new land. Lakes hold 44 times more of the planet's fresh water than rivers.

Thames
Although only 215 miles (346 km) long, it is England's most important waterway.

Loire
The longest river in France, it plays an important role in wine-growing in the area.

Mackenzie
The longest river in Canada, it freezes for about seven months of the year.

Great Lakes
The five lakes cover about 95,000 sq miles (245,000 sq km), which is an area larger than the entire UK.

Mississippi–Missouri
Supplies drinking water for more than 18 million people in the US.

Amazon
The Amazon basin (the area containing the river and all its tributaries) covers about 2.7 million sq miles (7 million sq km)—almost as large as Australia.

Madeira
The biggest tributary of the Amazon.

Congo
One of the deepest rivers in the world, with depths of more than 720 ft (220 m).

Paraná
The Paraná's Itaipú Dam is the second-largest hydroelectric power station in the world.

River volume

A river's size is measured by its discharge, or volume flow rate. This is the average amount of water that flows into the sea in a given time.

RIVER FLOW, IN MILLIONS OF GALLONS PER SECOND (MILLIONS OF LITERS PER SECOND)

- Amazon 57.9 (219)
- Congo–Chambeshi 11 (41.8)
- Yangtze 8.4 (31.9)
- Río de la Plata–Paraná–Paranaíba 5.8 (22)
- Yenisei–Angara–Selenga 5.2 (19.6)
- Mississippi–Missouri 4.4 (16.8)
- Ob–Irtysh 3.3 (12.5)
- Amur–Arqun 3 (11.4)
- Nile 0.7 (2.8)
- Yellow River 0.7 (2.6)

Longest rivers

Advances in GPS technology mean that scientists can now measure a river's length much more accurately than they could in the past.

LENGTH IN MILES   0   310
- Nile
- Amazon
- Yangtze
- Mississippi–Missouri
- Yenisei–Angara–Selenga
- Yellow River
- Ob–Irtysh
- Río de la Plata–Paraná–Paranaíba
- Congo–Chambeshi
- Amur–Arqun
LENGTH IN KILOMETERS   0   500

RIVERS ARE EARTH'S MOST POWERFUL FORCE OF EROSION, CARRYING

Ob' Ends in the Arctic Ocean.

Yenisey
Freezes along its entire length by mid-November each year.

Danube
Flows through 10 countries on its way to the Black Sea.

Lake Baikal
At about 25 million years old, by far the oldest lake on Earth.

Amur
Part of the Amur provides a natural boundary between Russia and the People's Republic of China.

Yellow River (Huang He)
So-called because of the huge amounts of mineral-rich silt it carries downstream.

Yangtze (Chang Jiang)
One of the world's busiest rivers, it flows through the major Chinese cities of Shanghai and Nanjing.

Seasonal rivers
Some rivers, shown in brown, flow only in the wet season. Some of these flow only in particularly wet years.

Nile
About 90 percent of the people of Egypt live close to the banks of the Nile.

Lake Victoria
The world's second-largest freshwater lake by area (after Lake Superior), it provides water for the Nile.

Ganges
Holy river to the world's 1.2 billion Hindus.

Murray–Darling
Makes up a large river basin in southeast Australia, connecting the Snowy Mountains to the Indian Ocean.

THE **AMAZON** CARRIES **ONE-FIFTH** OF ALL THE **FRESH WATER** EMPTIED INTO THE **OCEANS**

620   930   1,240   1,550   1,860   2,170   2,480   2,790   3,100   3,410   3,720   4,030

1,000   1,500   2,000   2,500   3,000   3,500   4,000   4,500   5,000   5,500   6,000   6,500

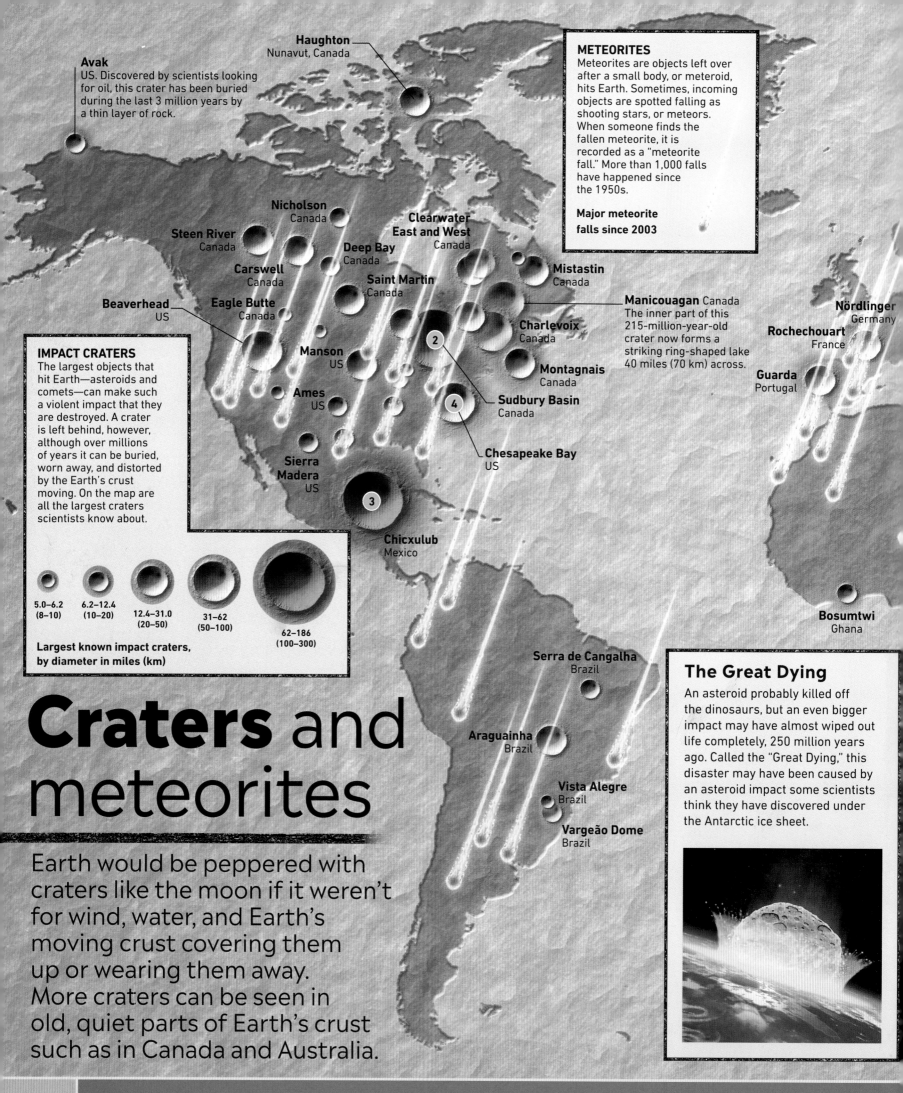

Avak
US. Discovered by scientists looking for oil, this crater has been buried during the last 3 million years by a thin layer of rock.

Haughton
Nunavut, Canada

METEORITES
Meteorites are objects left over after a small body, or meteroid, hits Earth. Sometimes, incoming objects are spotted falling as shooting stars, or meteors. When someone finds the fallen meteorite, it is recorded as a "meteorite fall." More than 1,000 falls have happened since the 1950s.

Major meteorite falls since 2003

Nicholson
Canada

Steen River
Canada

Clearwater East and West
Canada

Deep Bay
Canada

Carswell
Canada

Saint Martin
Canada

Mistastin
Canada

Beaverhead
US

Eagle Butte
Canada

Charlevoix
Canada

Manicouagan Canada
The inner part of this 215-million-year-old crater now forms a striking ring-shaped lake 40 miles (70 km) across.

Nördlinger
Germany

Rochechouart
France

IMPACT CRATERS
The largest objects that hit Earth—asteroids and comets—can make such a violent impact that they are destroyed. A crater is left behind, however, although over millions of years it can be buried, worn away, and distorted by the Earth's crust moving. On the map are all the largest craters scientists know about.

Manson
US

②

Montagnais
Canada

Guarda
Portugal

Ames
US

④

Sudbury Basin
Canada

Sierra Madera
US

Chesapeake Bay
US

③

5.0–6.2 (8–10)	6.2–12.4 (10–20)	12.4–31.0 (20–50)	31–62 (50–100)	62–186 (100–300)

Largest known impact craters, by diameter in miles (km)

Chicxulub
Mexico

Bosumtwi
Ghana

Craters and meteorites

Serra de Cangalha
Brazil

The Great Dying
An asteroid probably killed off the dinosaurs, but an even bigger impact may have almost wiped out life completely, 250 million years ago. Called the "Great Dying," this disaster may have been caused by an asteroid impact some scientists think they have discovered under the Antarctic ice sheet.

Araguainha
Brazil

Vista Alegre
Brazil

Vargeão Dome
Brazil

Earth would be peppered with craters like the moon if it weren't for wind, water, and Earth's moving crust covering them up or wearing them away. More craters can be seen in old, quiet parts of Earth's crust such as in Canada and Australia.

AN ASTEROID HITTING ANTARCTICA 250 MILLION YEARS AGO MAY HAVE

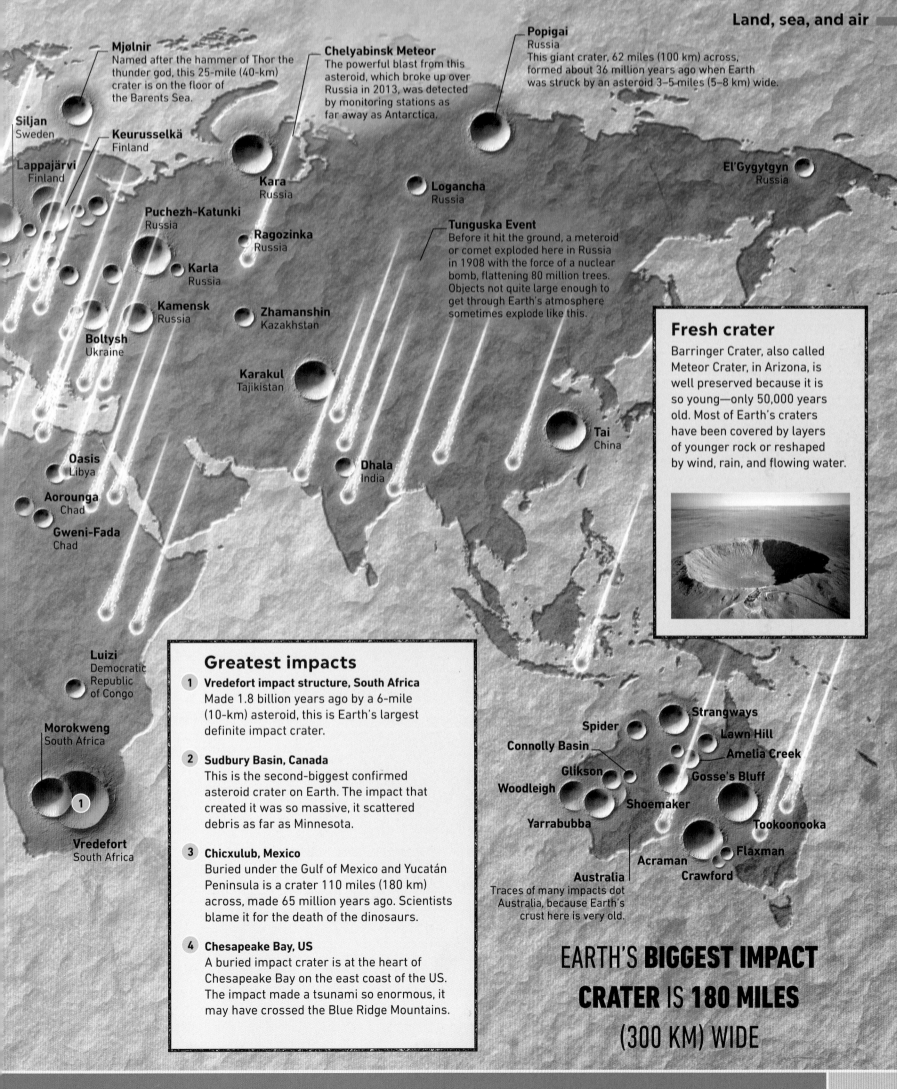

Mjølnir
Named after the hammer of Thor the thunder god, this 25-mile (40-km) crater is on the floor of the Barents Sea.

Siljan
Sweden

Keurusselkä
Finland

Lappajärvi
Finland

Chelyabinsk Meteor
The powerful blast from this asteroid, which broke up over Russia in 2013, was detected by monitoring stations as far away as Antarctica.

Popigai
Russia
This giant crater, 62 miles (100 km) across, formed about 36 million years ago when Earth was struck by an asteroid 3–5 miles (5–8 km) wide.

Kara
Russia

Logancha
Russia

El'Gygytgyn
Russia

Puchezh-Katunki
Russia

Ragozinka
Russia

Karla
Russia

Kamensk
Russia

Zhamanshin
Kazakhstan

Tunguska Event
Before it hit the ground, a meteroid or comet exploded here in Russia in 1908 with the force of a nuclear bomb, flattening 80 million trees. Objects not quite large enough to get through Earth's atmosphere sometimes explode like this.

Boltysh
Ukraine

Karakul
Tajikistan

Tai
China

Oasis
Libya

Dhala
India

Aorounga
Chad

Gweni-Fada
Chad

Luizi
Democratic Republic of Congo

Morokweng
South Africa

Vredefort
South Africa

Fresh crater

Barringer Crater, also called Meteor Crater, in Arizona, is well preserved because it is so young—only 50,000 years old. Most of Earth's craters have been covered by layers of younger rock or reshaped by wind, rain, and flowing water.

Greatest impacts

1. **Vredefort impact structure, South Africa**
Made 1.8 billion years ago by a 6-mile (10-km) asteroid, this is Earth's largest definite impact crater.

2. **Sudbury Basin, Canada**
This is the second-biggest confirmed asteroid crater on Earth. The impact that created it was so massive, it scattered debris as far as Minnesota.

3. **Chicxulub, Mexico**
Buried under the Gulf of Mexico and Yucatán Peninsula is a crater 110 miles (180 km) across, made 65 million years ago. Scientists blame it for the death of the dinosaurs.

4. **Chesapeake Bay, US**
A buried impact crater is at the heart of Chesapeake Bay on the east coast of the US. The impact made a tsunami so enormous, it may have crossed the Blue Ridge Mountains.

Spider

Strangways

Lawn Hill

Connolly Basin

Amelia Creek

Glikson

Gosse's Bluff

Woodleigh

Shoemaker

Yarrabubba

Tookoonooka

Acraman

Flaxman

Crawford

Australia
Traces of many impacts dot Australia, because Earth's crust here is very old.

EARTH'S BIGGEST IMPACT CRATER IS 180 MILES (300 KM) WIDE

Prospect Creek, US
At −80°F (−62.2°C), this is the seventh-coldest place on the planet.

Malgovik, Sweden
The coldest spot in Sweden, with a record of −63.4°F (−53°C).

Snag, Canada
Recorded temperature of −81°F (−63°C), to make it the coldest site in North America.

Klinck Automated Weather Station, Greenland
Fourth-coldest spot on Earth, at −93.3°F (−69.6°C).

Furnace Creek, US
The world's highest ever air temperature, 134°F (56.7°C), was recorded here in 1913.

Kebili, Tunisia
Recorded temperature of 131°F (55°C) in 1931, tying for the third-hottest place ever.

Mexicali area, Mexico
A 1995 record of 125.6 °F (52.0°C).

Daily differences

Many deserts are hot during the day but drastically cooler at night. With no clouds or mist in the way of the sun, the ground warms up fast during the day. With no blanketing cloud at night, the heat escapes quickly. In humid climates, daily temperatures vary a lot less.

Cold mountains
The higher up you are, the lower the air pressure—and the temperature. The Andes mountain range is much colder than the land that surrounds it.

Al 'Aziziyah, Libya
Lost its title as world's hottest place in 2012, when weather scientists found its 1922 record measurement was probably wrong.

(1) **Luxor, Egypt**
Luxor has a dry, desert climate. In June, the daily temperature varies greatly, from an average maximum of 105.8°F (41°C) down to 71.6°F (22°C) at night.

(2) **Singapore**
Singapore's climate is very warm and humid all year round. In June, the daily temperature varies from 88.3°F (31.3°C) to a sticky 76.5°F (24.7°C) at night.

Blazing summers, freezing winters

In the middle of large continents, it is often hot in summer and very cold in winter. In coastal areas, warm or cool winds and currents carried by the sea moderate temperatures. Without this balance, inland areas can become extremely hot or cold.

(1) **Verkhoyansk, Russia**
The world's biggest seasonal temperature differences are found in Verkhoyansk. The highest temperature ever recorded was 103.8°F (39.9°C) and the lowest was −90°F (−67.8°C).

(2) **Regina, Canada**
Regina's highest-ever temperature was 109.9°F (43.3°C) and the lowest was -58°F (-50°C).

IN **1924**, THE AUSTRALIAN TOWN OF **MARBLE BAR** REACHED 100°F (37.8°C) OR ABOVE FOR **160 DAYS** IN A ROW

Amundsen–Scott Station, South Pole
The second-coldest point on Earth, at −117°F (−82.8°C).

THE "GULF STREAM" OCEAN CURRENT FROM THE GULF OF MEXICO GIVES

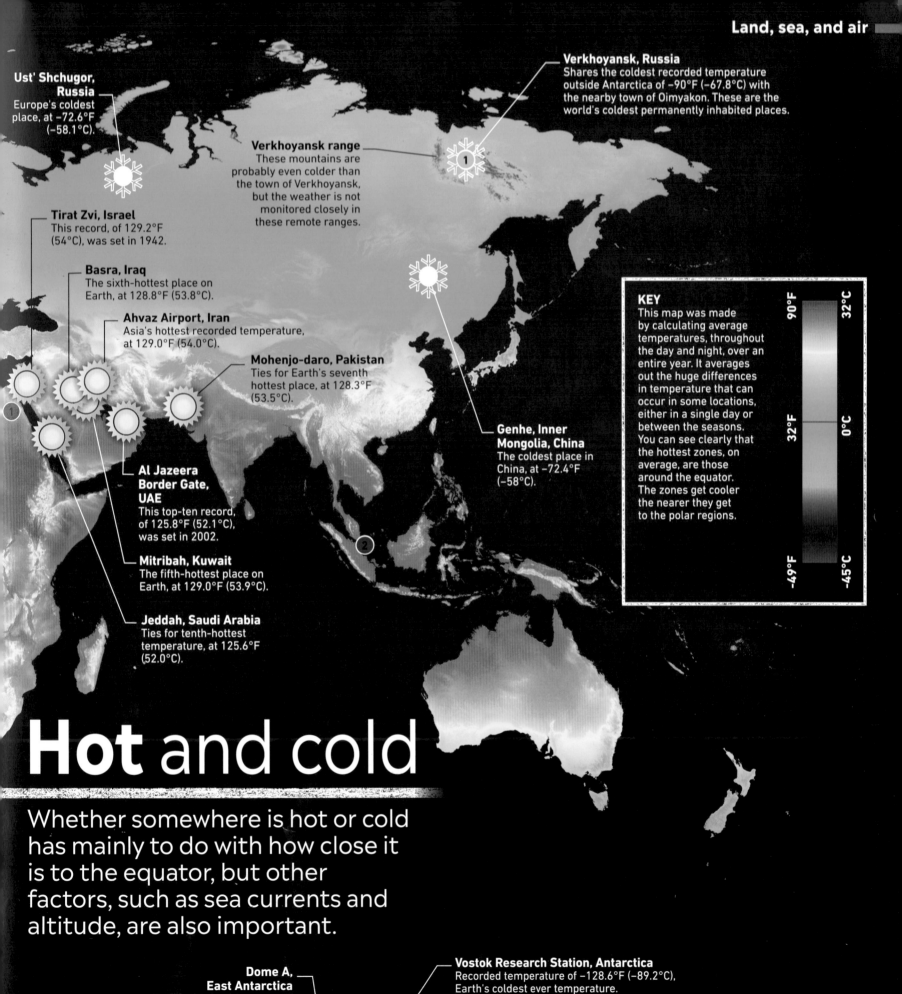

Ust' Shchugor, Russia
Europe's coldest place, at −72.6°F (−58.1°C).

Verkhoyansk range
These mountains are probably even colder than the town of Verkhoyansk, but the weather is not monitored closely in these remote ranges.

Verkhoyansk, Russia
Shares the coldest recorded temperature outside Antarctica of −90°F (−67.8°C) with the nearby town of Oimyakon. These are the world's coldest permanently inhabited places.

Tirat Zvi, Israel
This record, of 129.2°F (54°C), was set in 1942.

Basra, Iraq
The sixth-hottest place on Earth, at 128.8°F (53.8°C).

Ahvaz Airport, Iran
Asia's hottest recorded temperature, at 129.0°F (54.0°C).

Mohenjo-daro, Pakistan
Ties for Earth's seventh hottest place, at 128.3°F (53.5°C).

Genhe, Inner Mongolia, China
The coldest place in China, at −72.4°F (−58°C).

Al Jazeera Border Gate, UAE
This top-ten record, of 125.8°F (52.1°C), was set in 2002.

Mitribah, Kuwait
The fifth-hottest place on Earth, at 129.0°F (53.9°C).

Jeddah, Saudi Arabia
Ties for tenth-hottest temperature, at 125.6°F (52.0°C).

KEY
This map was made by calculating average temperatures, throughout the day and night, over an entire year. It averages out the huge differences in temperature that can occur in some locations, either in a single day or between the seasons. You can see clearly that the hottest zones, on average, are those around the equator. The zones get cooler the nearer they get to the polar regions.

90°F — 32°C
32°F — 0°C
−49°F — −45°C

Hot and cold

Whether somewhere is hot or cold has mainly to do with how close it is to the equator, but other factors, such as sea currents and altitude, are also important.

Dome A, East Antarctica
Earth's third-coldest spot, at −116.5 °F (−82.5°C).

Vostok Research Station, Antarctica
Recorded temperature of −128.6°F (−89.2°C), Earth's coldest ever temperature.

NORTHWEST EUROPE WARMER WINTERS THAN PLACES FARTHER SOUTH.

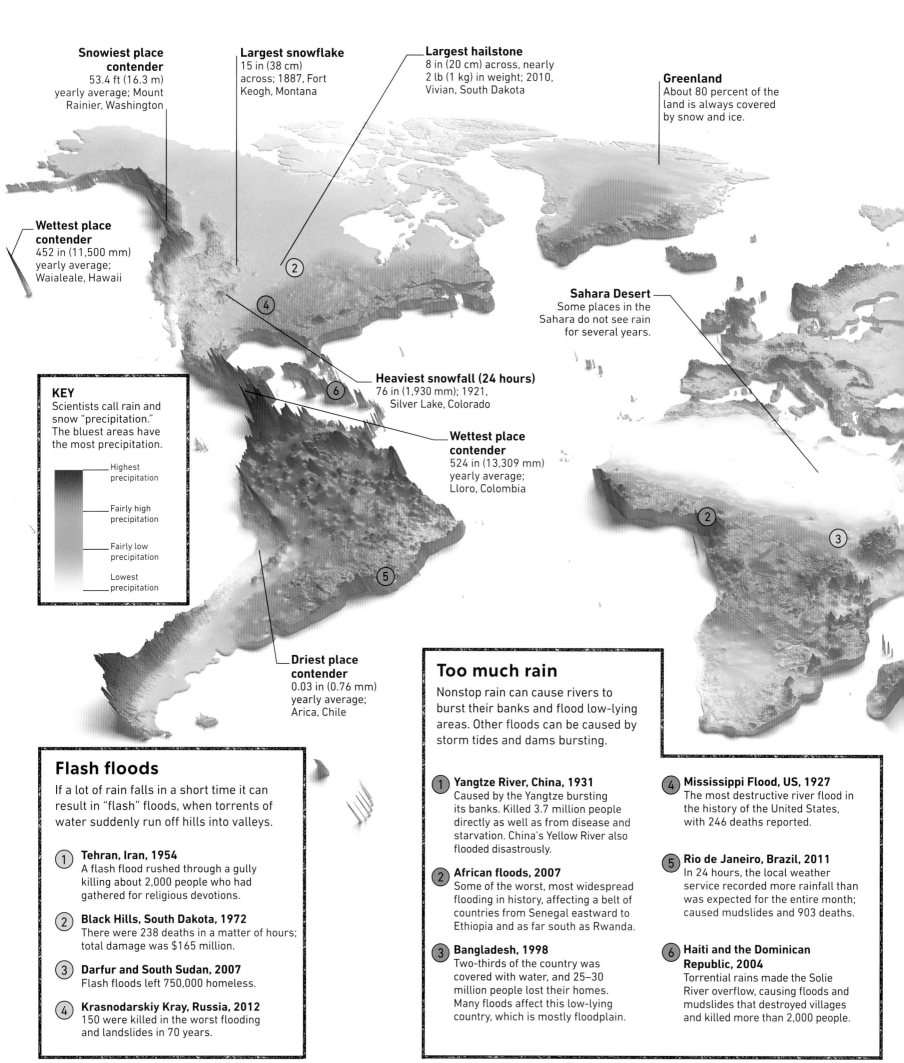

Snowiest place contender
53.4 ft (16.3 m) yearly average; Mount Rainier, Washington

Largest snowflake
15 in (38 cm) across; 1887, Fort Keogh, Montana

Largest hailstone
8 in (20 cm) across, nearly 2 lb (1 kg) in weight; 2010, Vivian, South Dakota

Greenland
About 80 percent of the land is always covered by snow and ice.

Wettest place contender
452 in (11,500 mm) yearly average; Waialeale, Hawaii

Sahara Desert
Some places in the Sahara do not see rain for several years.

KEY
Scientists call rain and snow "precipitation." The bluest areas have the most precipitation.

- Highest precipitation
- Fairly high precipitation
- Fairly low precipitation
- Lowest precipitation

Heaviest snowfall (24 hours)
76 in (1,930 mm); 1921, Silver Lake, Colorado

Wettest place contender
524 in (13,309 mm) yearly average; Lloro, Colombia

Driest place contender
0.03 in (0.76 mm) yearly average; Arica, Chile

Flash floods

If a lot of rain falls in a short time it can result in "flash" floods, when torrents of water suddenly run off hills into valleys.

1. **Tehran, Iran, 1954**
A flash flood rushed through a gully killing about 2,000 people who had gathered for religious devotions.

2. **Black Hills, South Dakota, 1972**
There were 238 deaths in a matter of hours; total damage was $165 million.

3. **Darfur and South Sudan, 2007**
Flash floods left 750,000 homeless.

4. **Krasnodarskiy Kray, Russia, 2012**
150 were killed in the worst flooding and landslides in 70 years.

Too much rain

Nonstop rain can cause rivers to burst their banks and flood low-lying areas. Other floods can be caused by storm tides and dams bursting.

1. **Yangtze River, China, 1931**
Caused by the Yangtze bursting its banks. Killed 3.7 million people directly as well as from disease and starvation. China's Yellow River also flooded disastrously.

2. **African floods, 2007**
Some of the worst, most widespread flooding in history, affecting a belt of countries from Senegal eastward to Ethiopia and as far south as Rwanda.

3. **Bangladesh, 1998**
Two-thirds of the country was covered with water, and 25–30 million people lost their homes. Many floods affect this low-lying country, which is mostly floodplain.

4. **Mississippi Flood, US, 1927**
The most destructive river flood in the history of the United States, with 246 deaths reported.

5. **Rio de Janeiro, Brazil, 2011**
In 24 hours, the local weather service recorded more rainfall than was expected for the entire month; caused mudslides and 903 deaths.

6. **Haiti and the Dominican Republic, 2004**
Torrential rains made the Solie River overflow, causing floods and mudslides that destroyed villages and killed more than 2,000 people.

Rain and snow

197 IN (5,000 MM) OF RAIN MAY FALL IN ONE PLACE DURING INDIA'S **MONSOON** SEASON

Rainfall varies dramatically with place. Torrential rain drenches southern Asia during the monsoon season, yet some desert regions have virtually no rain at all. Near the poles, very little snow falls, but the snow rarely melts, so some land is permanently under a layer of ice.

④
①

Heaviest rainfall (1 month, and 1 year)
370 in (9,300 mm) and 905 in (22,987 mm); both 1860–61, Cherrapunji, India

①

③

Arabian Peninsula
As in the Sahara, there is very little rain in this largely desert region.

Heaviest rainfall (24 hours)
71.9 in (1,825 mm); 1966, Foc-Foc, Réunion, during Tropical Cyclone Denise

Snowiest place contender
49.5 ft (15 m) yearly average; Niseko, Japan

Borneo
Many equatorial rainforests, such as those in Borneo, have no dry season, and it rains every day.

Monsoon extremes

Chittagong, in Bangladesh, has almost no rain in the dry season, but its monsoon rains are torrential. Paris, in France, has much more even monthly rainfall.

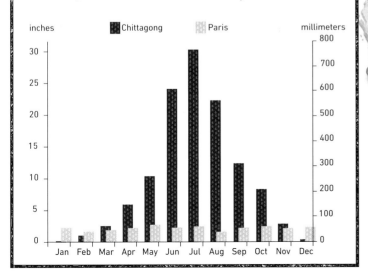

| inches | ■ Chittagong | ▫ Paris | millimeters |

Australia
This is the driest inhabited continent.

New Zealand
Rainfall is fairly high and is spread evenly throughout the year.

Driest place on Earth
0 in (0 mm) yearly average; Antarctica's Dry Valleys, which are free of snow and ice.

▼

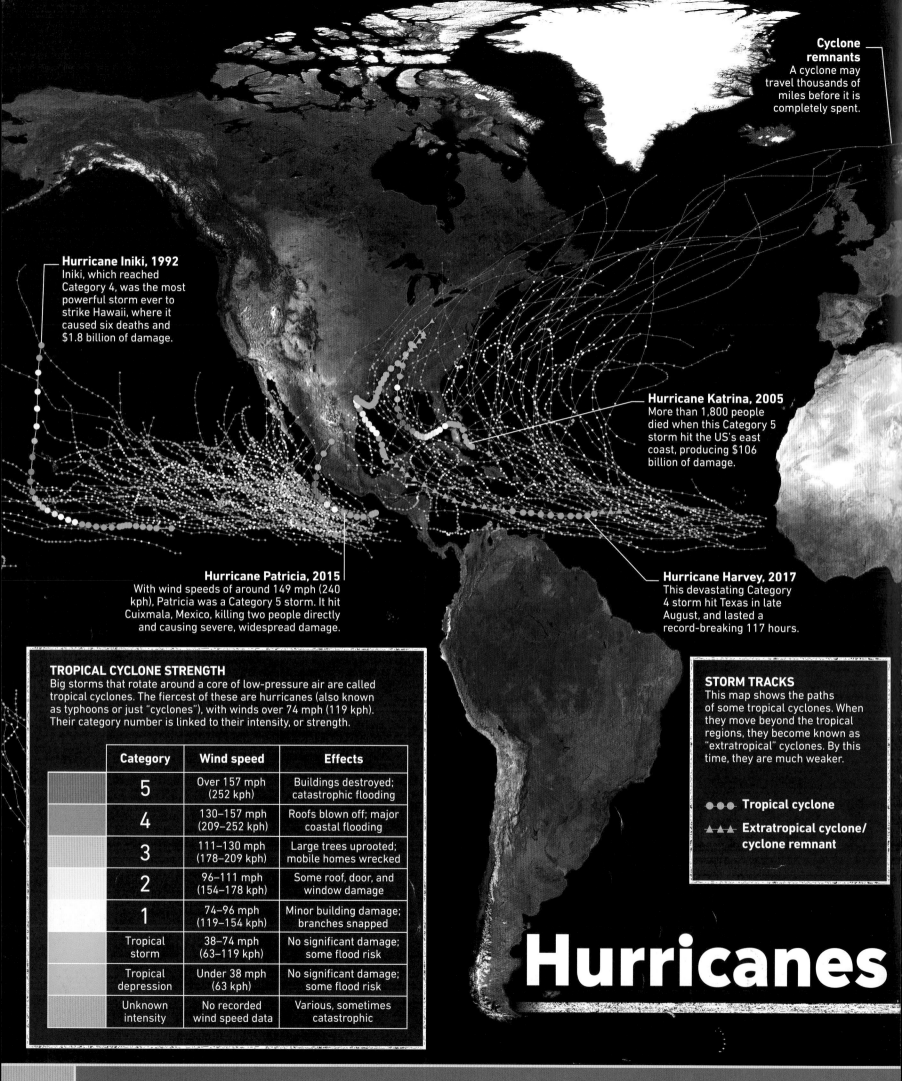

Cyclone remnants
A cyclone may travel thousands of miles before it is completely spent.

Hurricane Iniki, 1992
Iniki, which reached Category 4, was the most powerful storm ever to strike Hawaii, where it caused six deaths and $1.8 billion of damage.

Hurricane Katrina, 2005
More than 1,800 people died when this Category 5 storm hit the US's east coast, producing $106 billion of damage.

Hurricane Patricia, 2015
With wind speeds of around 149 mph (240 kph), Patricia was a Category 5 storm. It hit Cuixmala, Mexico, killing two people directly and causing severe, widespread damage.

Hurricane Harvey, 2017
This devastating Category 4 storm hit Texas in late August, and lasted a record-breaking 117 hours.

TROPICAL CYCLONE STRENGTH
Big storms that rotate around a core of low-pressure air are called tropical cyclones. The fiercest of these are hurricanes (also known as typhoons or just "cyclones"), with winds over 74 mph (119 kph). Their category number is linked to their intensity, or strength.

Category	Wind speed	Effects
5	Over 157 mph (252 kph)	Buildings destroyed; catastrophic flooding
4	130–157 mph (209–252 kph)	Roofs blown off; major coastal flooding
3	111–130 mph (178–209 kph)	Large trees uprooted; mobile homes wrecked
2	96–111 mph (154–178 kph)	Some roof, door, and window damage
1	74–96 mph (119–154 kph)	Minor building damage; branches snapped
Tropical storm	38–74 mph (63–119 kph)	No significant damage; some flood risk
Tropical depression	Under 38 mph (63 kph)	No significant damage; some flood risk
Unknown intensity	No recorded wind speed data	Various, sometimes catastrophic

STORM TRACKS
This map shows the paths of some tropical cyclones. When they move beyond the tropical regions, they become known as "extratropical" cyclones. By this time, they are much weaker.

●●● Tropical cyclone

▲▲▲ Extratropical cyclone/ cyclone remnant

Hurricanes

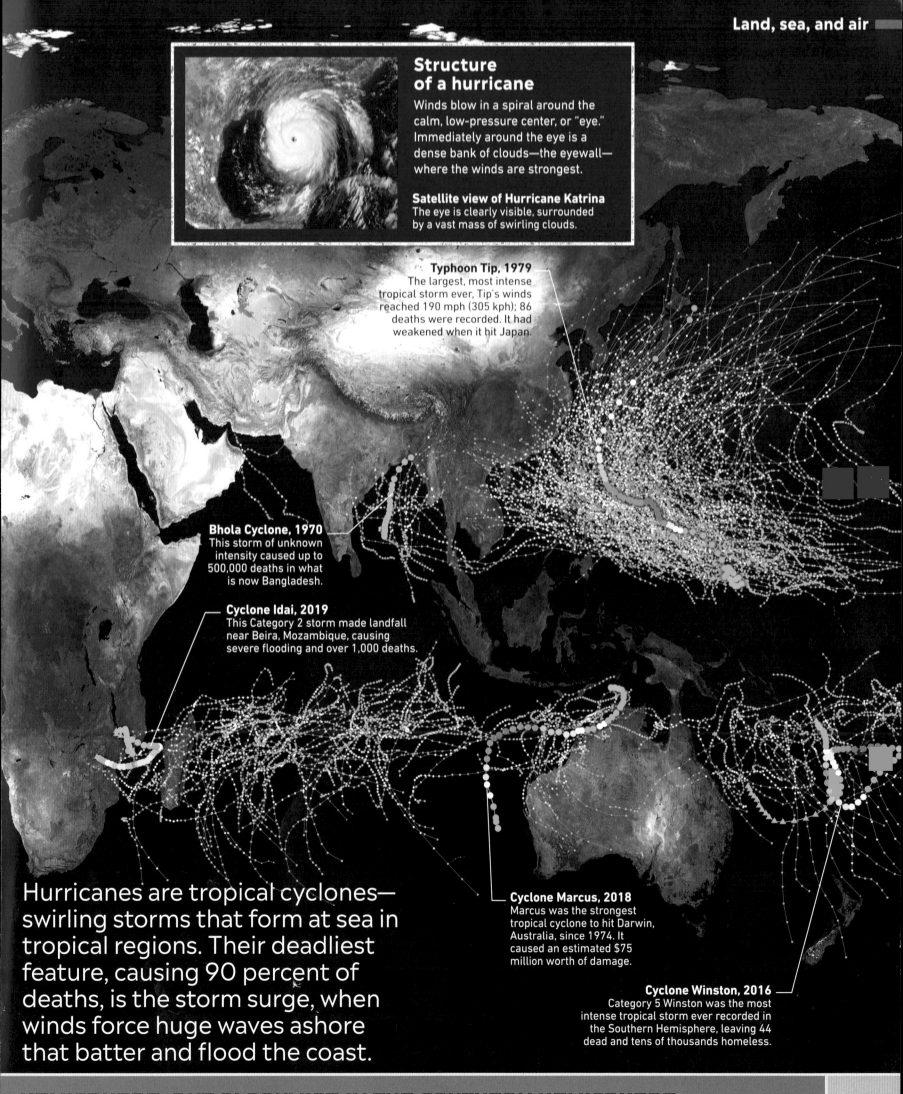

Structure of a hurricane

Winds blow in a spiral around the calm, low-pressure center, or "eye." Immediately around the eye is a dense bank of clouds—the eyewall—where the winds are strongest.

Satellite view of Hurricane Katrina
The eye is clearly visible, surrounded by a vast mass of swirling clouds.

Typhoon Tip, 1979
The largest, most intense tropical storm ever, Tip's winds reached 190 mph (305 kph); 86 deaths were recorded. It had weakened when it hit Japan.

Bhola Cyclone, 1970
This storm of unknown intensity caused up to 500,000 deaths in what is now Bangladesh.

Cyclone Idai, 2019
This Category 2 storm made landfall near Beira, Mozambique, causing severe flooding and over 1,000 deaths.

Cyclone Marcus, 2018
Marcus was the strongest tropical cyclone to hit Darwin, Australia, since 1974. It caused an estimated $75 million worth of damage.

Cyclone Winston, 2016
Category 5 Winston was the most intense tropical storm ever recorded in the Southern Hemisphere, leaving 44 dead and tens of thousands homeless.

Hurricanes are tropical cyclones—swirling storms that form at sea in tropical regions. Their deadliest feature, causing 90 percent of deaths, is the storm surge, when winds force huge waves ashore that batter and flood the coast.

Tropical broad-leaved moist forest
Also known as rainforest, these warm, wet woods support a huge variety of animal and plant life.

Tropical broad-leaved dry forest
These areas are warm all year round but have a long dry season, and many trees lose their leaves.

Tropical coniferous forest
Many migrating birds and butterflies spend the winter in these warm, dense conifer forests.

Temperate broad-leaved forest
The most common habitat of northern Europe and home to trees that lose their leaves in winter.

Temperate coniferous forest
Giant trees, such as the California redwood, thrive in these regions of warm summers and cool winters.

Boreal forest
Also called taiga, this is the largest land biome on Earth. It is dominated by just a few types of coniferous trees.

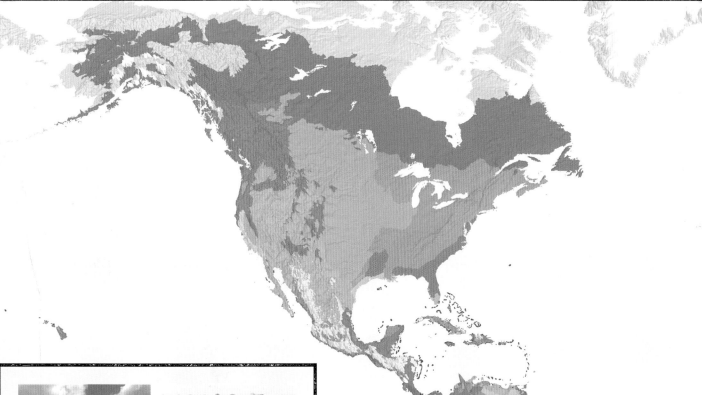

Savanna
A long dry season and short rainy periods results in a grassland studded with trees and herds of grazing animals.

Flooded savanna
Birds are attracted to these marshy wetland areas that are flooded in the wet season but grassland at other times.

Temperate grassland
Also known as prairie, steppe, or pampas, many of these vast, fertile plains are now farmland.

Mountain grassland
The inhabitants of these remote, high habitats must adapt to the cold and the intense sunlight.

Coral reef
The warm, shallow waters of a reef support a huge variety of life, from sharks to tiny sea horses.

Marine Biomes

Sea biomes are as varied as those on land. From beaches to the darkest ocean depths, living things find ways to survive and thrive.

Mangrove
On the shore, the mangroves' thick, tangled roots slow the water's flow and create a swamp.

TROPICAL RAINFORESTS COVER ABOUT 6 PERCENT OF THE LAND,

Mediterranean shrubland
Hot, dry summers can lead to fires that actually help the biome's typical shrubby plants sprout.

Desert and dry shrubland
Desert inhabitants have to be able to survive on less than 10 in (250 mm) of rainfall per year.

Arctic tundra
A cold, dry biome where the soil stays frozen at depth. This permafrost stops trees from growing.

Polar desert
Too cold and dry for almost all plants. Only animals dependent on the sea, such as penguins, can live here.

A **BIOME'S PLANTS** AND **ANIMALS** FORM A **COMPLEX** AND **INTERCONNECTED** COMMUNITY

Biomes

A biome is an area that we define according to the animals and plants that live there. They have to adapt to the biome's specific conditions such as temperature, type of soil, and the amount of light and water.

BUT ARE HOME TO NEARLY HALF OF ALL LIVING THINGS ON EARTH.

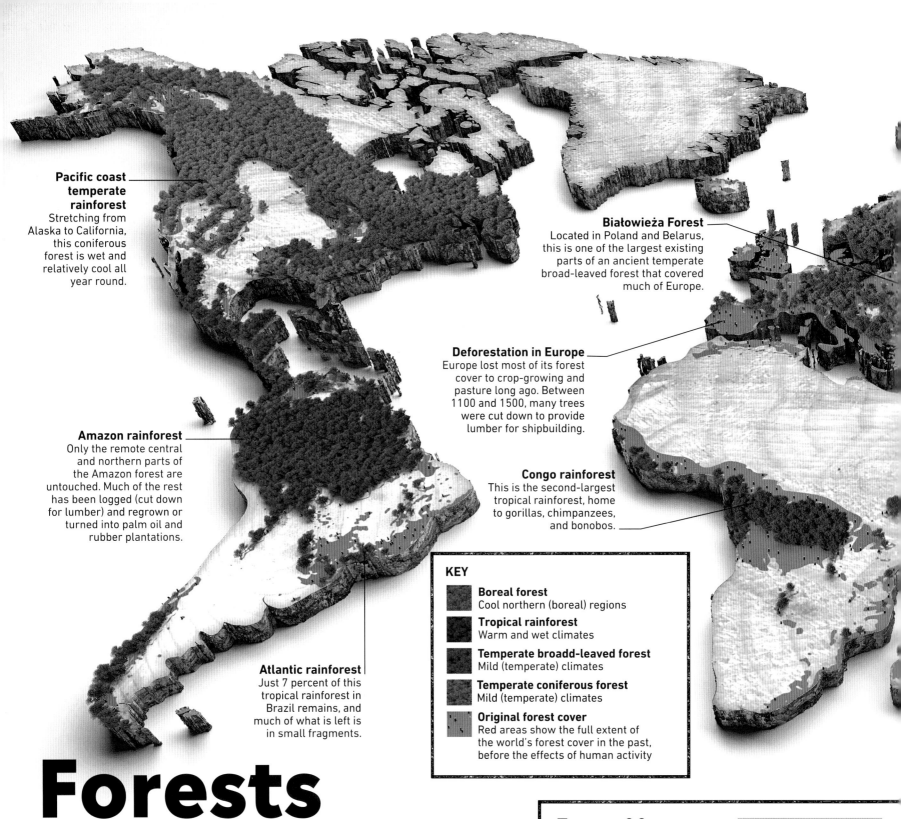

Pacific coast temperate rainforest
Stretching from Alaska to California, this coniferous forest is wet and relatively cool all year round.

Białowieża Forest
Located in Poland and Belarus, this is one of the largest existing parts of an ancient temperate broad-leaved forest that covered much of Europe.

Deforestation in Europe
Europe lost most of its forest cover to crop-growing and pasture long ago. Between 1100 and 1500, many trees were cut down to provide lumber for shipbuilding.

Amazon rainforest
Only the remote central and northern parts of the Amazon forest are untouched. Much of the rest has been logged (cut down for lumber) and regrown or turned into palm oil and rubber plantations.

Congo rainforest
This is the second-largest tropical rainforest, home to gorillas, chimpanzees, and bonobos.

Atlantic rainforest
Just 7 percent of this tropical rainforest in Brazil remains, and much of what is left is in small fragments.

KEY

- **Boreal forest**
  Cool northern (boreal) regions
- **Tropical rainforest**
  Warm and wet climates
- **Temperate broadd-leaved forest**
  Mild (temperate) climates
- **Temperate coniferous forest**
  Mild (temperate) climates
- **Original forest cover**
  Red areas show the full extent of the world's forest cover in the past, before the effects of human activity

Forests

Forests are vital to life on Earth. They make the air breathable, protect the soil, and preserve fresh water supplies. But they are disappearing— and while efforts are being made to slow deforestation, about 25 million acres are still lost each year.

Types of forests

Forests differ according to climate. Each type of forest has its own distinct collection of trees, forest-floor plants, and animal life. Tropical rainforests are the most diverse—30 percent of all plant and animal species live in the Amazon alone. Some tropical forests are evergreen, while in others the trees lose their leaves in the dry season.

Temperate broad-leaved
Deciduous trees, such as oak and beech. Herbs, ferns, and shrubs on the forest floor.

Taiga

This vast belt of boreal forest stretches right across northern Europe and Asia. In the east, it is wilderness, but much in the west is working forest, managed for lumber and paper production.

Disappearing forests

With the world's population growing, demand for lumber and land for farming and towns has increased the rate of forest clearance. Here you can see the decline in Borneo's forests from 1950 to 2010.

Borneo, 1950   **1985**   **2010**

AT CURRENT RATES OF **LOGGING,** IN **100 YEARS** WE WILL **NO LONGER** HAVE ANY **RAINFORESTS**

Japan

Japan retains a lot of its original woodland and is the most thickly forested industrialized country.

Borneo

Home of most of the world's orangutans, Borneo's rainforest has declined by more than 50 percent since the mid-20th century (see above).

New Guinea

Two-thirds of New Guinea is largely unspoiled rainforest, with many unique species. It is at risk from logging, mining, and agriculture.

Tropical rainforest

As many as 300 tree species per 2.5 acres (hectare). Often rich in forest-floor plants.

Boreal forest

Hardy conifers, such as larch, spruce, fir, and pine. Mosses dominate the forest floor.

Australia

About 38 percent of Australia's forests have been lost since European settlers arrived around 200 years ago.

New Zealand

The remote southwest of New Zealand is home to unique temperate rainforests full of lush tree ferns.

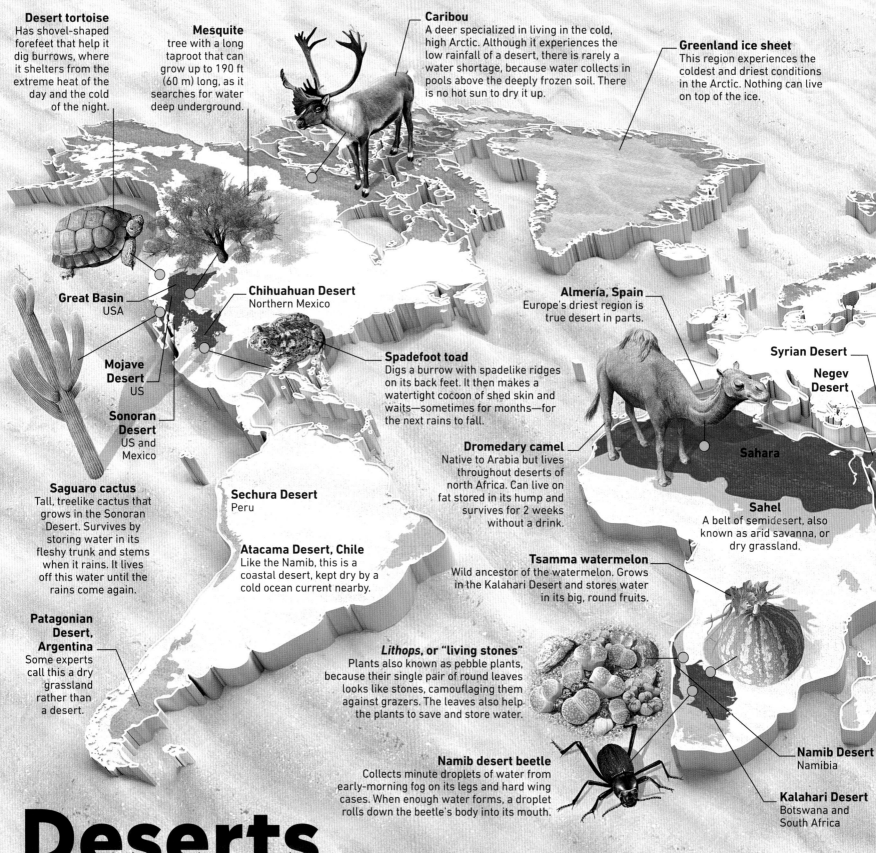

Desert tortoise
Has shovel-shaped forefeet that help it dig burrows, where it shelters from the extreme heat of the day and the cold of the night.

Mesquite
tree with a long taproot that can grow up to 190 ft (60 m) long, as it searches for water deep underground.

Caribou
A deer specialized in living in the cold, high Arctic. Although it experiences the low rainfall of a desert, there is rarely a water shortage, because water collects in pools above the deeply frozen soil. There is no hot sun to dry it up.

Greenland ice sheet
This region experiences the coldest and driest conditions in the Arctic. Nothing can live on top of the ice.

Great Basin
USA

Chihuahuan Desert
Northern Mexico

Almería, Spain
Europe's driest region is true desert in parts.

Syrian Desert

Negev Desert

Mojave Desert
US

Spadefoot toad
Digs a burrow with spadelike ridges on its back feet. It then makes a watertight cocoon of shed skin and waits—sometimes for months—for the next rains to fall.

Sonoran Desert
US and Mexico

Sahara

Dromedary camel
Native to Arabia but lives throughout deserts of north Africa. Can live on fat stored in its hump and survives for 2 weeks without a drink.

Saguaro cactus
Tall, treelike cactus that grows in the Sonoran Desert. Survives by storing water in its fleshy trunk and stems when it rains. It lives off this water until the rains come again.

Sechura Desert
Peru

Sahel
A belt of semidesert, also known as arid savanna, or dry grassland.

Atacama Desert, Chile
Like the Namib, this is a coastal desert, kept dry by a cold ocean current nearby.

Tsamma watermelon
Wild ancestor of the watermelon. Grows in the Kalahari Desert and stores water in its big, round fruits.

Patagonian Desert, Argentina
Some experts call this a dry grassland rather than a desert.

***Lithops,* or "living stones"**
Plants also known as pebble plants, because their single pair of round leaves looks like stones, camouflaging them against grazers. The leaves also help the plants to save and store water.

Namib desert beetle
Collects minute droplets of water from early-morning fog on its legs and hard wing cases. When enough water forms, a droplet rolls down the beetle's body into its mouth.

Namib Desert
Namibia

Kalahari Desert
Botswana and South Africa

Deserts

Deserts are found from the icy poles to the tropics. So while all deserts have low rainfall—less than 10 in (250 mm) a year, and often much less—they are not always hot. Even in hot deserts, the nights are often cold.

Antarctica
One of the most arid parts of Earth's largest desert is its Dry Valleys region (right), the only area of Antarctica not covered in thick ice, and where there is almost no snowfall. Cold, dry winds blast down from mountain peaks and turn all moisture to water vapor.

Desert terrain

Deserts range widely in how they look. Soil forms very slowly and the land is often bare rock or gravel. Any loose, sandy soil may be blown into dunes. Sometimes, though, tough grasses or fleshy plants bind the soil together.

Dunes, or "sand seas"
Shifting mountains of sand can prevent plant growth.

Rock and gravel
Where no plants grow, the bedrock is often visible.

Dry grassland
Desert grasses can form soil and provide food for grazers.

Fleshy plants
Fleshy, water-storing plants may form thick vegetation.

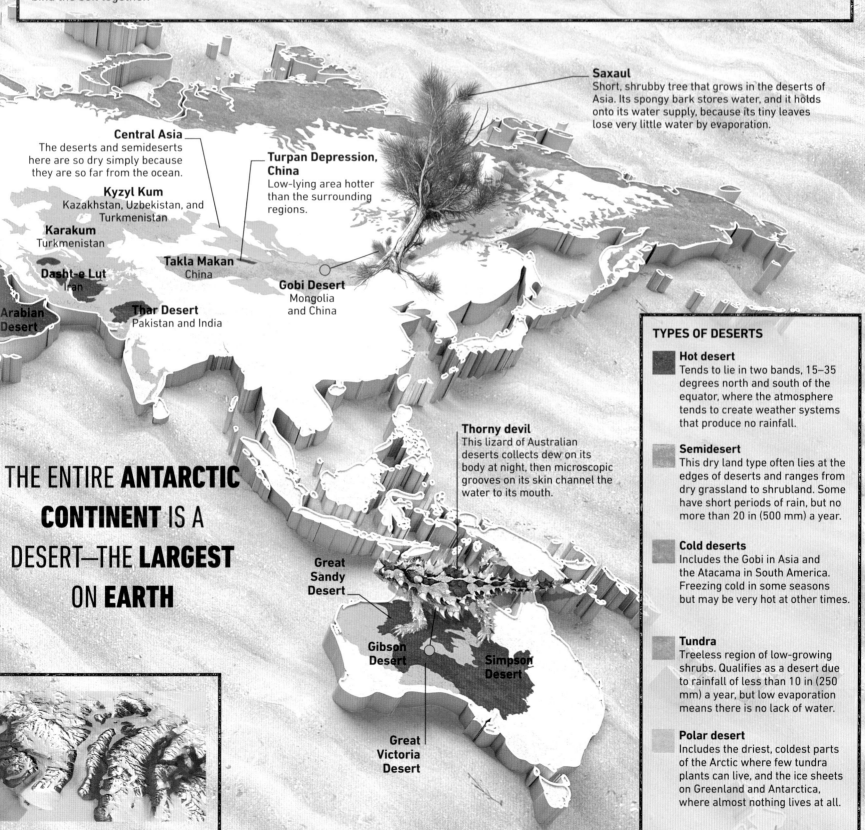

Saxaul
Short, shrubby tree that grows in the deserts of Asia. Its spongy bark stores water, and it holds onto its water supply, because its tiny leaves lose very little water by evaporation.

Central Asia
The deserts and semideserts here are so dry simply because they are so far from the ocean.

Turpan Depression, China
Low-lying area hotter than the surrounding regions.

Kyzyl Kum
Kazakhstan, Uzbekistan, and Turkmenistan

Karakum
Turkmenistan

Dasht-e Lut
Iran

Takla Makan
China

Arabian Desert

Thar Desert
Pakistan and India

Gobi Desert
Mongolia and China

Thorny devil
This lizard of Australian deserts collects dew on its body at night, then microscopic grooves on its skin channel the water to its mouth.

THE ENTIRE **ANTARCTIC CONTINENT** IS A DESERT—THE **LARGEST** ON **EARTH**

Great Sandy Desert

Gibson Desert

Simpson Desert

Great Victoria Desert

TYPES OF DESERTS

Hot desert
Tends to lie in two bands, 15–35 degrees north and south of the equator, where the atmosphere tends to create weather systems that produce no rainfall.

Semidesert
This dry land type often lies at the edges of deserts and ranges from dry grassland to shrubland. Some have short periods of rain, but no more than 20 in (500 mm) a year.

Cold deserts
Includes the Gobi in Asia and the Atacama in South America. Freezing cold in some seasons but may be very hot at other times.

Tundra
Treeless region of low-growing shrubs. Qualifies as a desert due to rainfall of less than 10 in (250 mm) a year, but low evaporation means there is no lack of water.

Polar desert
Includes the driest, coldest parts of the Arctic where few tundra plants can live, and the ice sheets on Greenland and Antarctica, where almost nothing lives at all.

Ice

Ice covers one-tenth of Earth's land surface, mostly in the polar regions. At earlier times in Earth's history, when the climate was much cooler, ice covered an area up to three times larger than it does today.

Sea ice

Sea ice is frozen sea. It forms when the ocean's surface freezes in winter. Where it lasts year round, it may be 20 ft (6 m) thick—elsewhere it is thinner. "Pancake ice" (right) is disks of sea ice up to 4 in (10 cm) thick.

Summer ice The polar sea ice cover shrinks in summer, but some sea always remains under a layer of ice.

Winter ice As the weather gets colder, the polar sea ice spreads far beyond its summer limits.

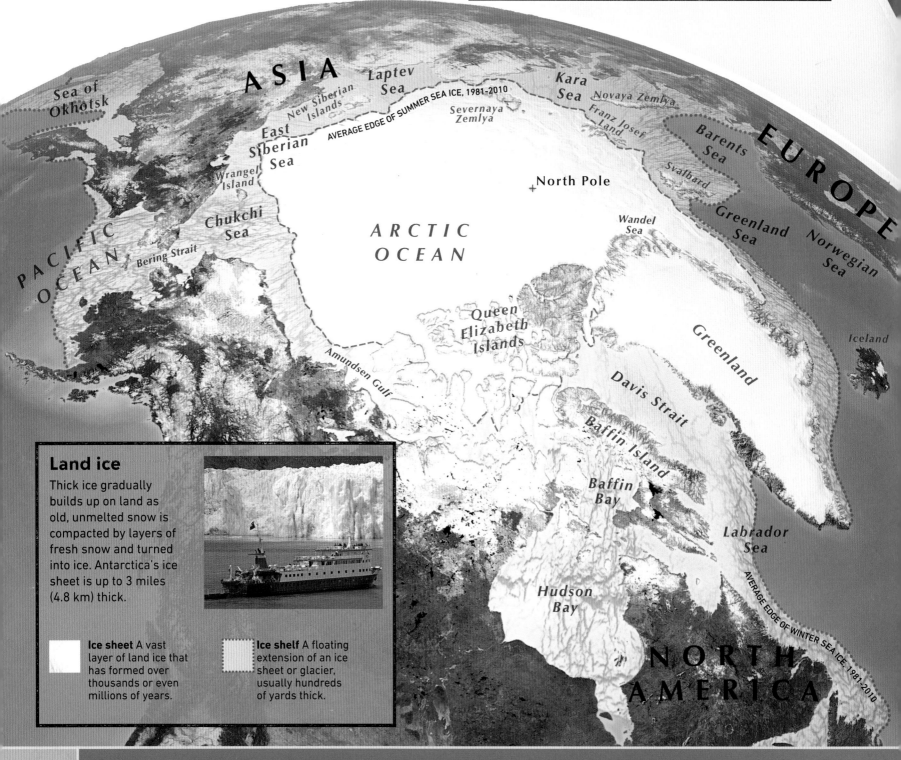

SOUTH AMERICA

ASIA

Sea of Okhotsk

New Siberian Islands

Laptev Sea

AVERAGE EDGE OF SUMMER SEA ICE, 1981-2010

Severnaya Zemlya

Kara Sea

Novaya Zemlya

Franz Josef Land

Barents Sea

EUROPE

East Siberian Sea

Wrangel Island

Svalbard

Chukchi Sea

North Pole

ARCTIC OCEAN

Wandel Sea

Greenland Sea

Norwegian Sea

PACIFIC OCEAN

Bering Strait

Amundsen Gulf

Queen Elizabeth Islands

Greenland

Iceland

Davis Strait

Baffin Island

Baffin Bay

Labrador Sea

Hudson Bay

AVERAGE EDGE OF WINTER SEA ICE 1981-2010

NORTH AMERICA

Land ice

Thick ice gradually builds up on land as old, unmelted snow is compacted by layers of fresh snow and turned into ice. Antarctica's ice sheet is up to 3 miles (4.8 km) thick.

Ice sheet A vast layer of land ice that has formed over thousands or even millions of years.

Ice shelf A floating extension of an ice sheet or glacier, usually hundreds of yards thick.

THE TALLEST ICEBERG EVER SEEN IN THE NORTH ATLANTIC ROSE 551 FT

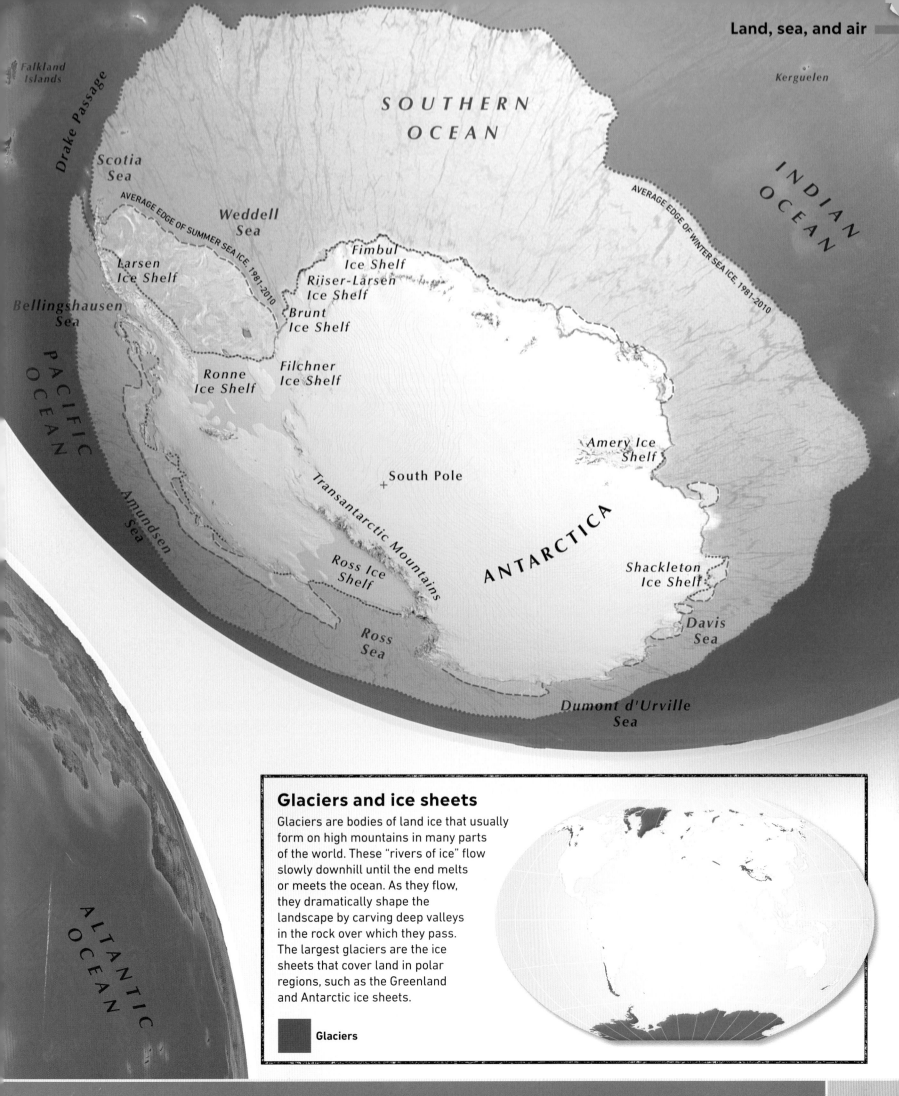

Falkland
Islands

Kerguelen

S O U T H E R N
O C E A N

I
N
D
I
A
N

O
C
E
A
N

Drake Passage

Scotia
Sea

AVERAGE EDGE OF SUMMER SEA ICE 1981-2010

AVERAGE EDGE OF WINTER SEA ICE 1981-2010

Weddell
Sea

Fimbul
Ice Shelf

Larsen
Ice Shelf

Riiser-Larsen
Ice Shelf

Bellingshausen
Sea

Brunt
Ice Shelf

P
A
C
I
F
I
C

O
C
E
A
N

Ronne
Ice Shelf

Filchner
Ice Shelf

Amery Ice
Shelf

+ South Pole

Transantarctic Mountains

Amundsen
Sea

ANTARCTICA

Shackleton
Ice Shelf

Ross Ice
Shelf

Davis
Sea

Ross
Sea

Dumont d'Urville
Sea

A
T
L
A
N
T
I
C

O
C
E
A
N

Glaciers and ice sheets

Glaciers are bodies of land ice that usually
form on high mountains in many parts
of the world. These "rivers of ice" flow
slowly downhill until the end melts
or meets the ocean. As they flow,
they dramatically shape the
landscape by carving deep valleys
in the rock over which they pass.
The largest glaciers are the ice
sheets that cover land in polar
regions, such as the Greenland
and Antarctic ice sheets.

█ Glaciers

Time zones map

The map shows the time of day at 12 noon Coordinated Universal Time (UTC), the base from which all times are set. The columns are time zones labeled with the number of hours they are ahead or behind UTC. If you stood halfway between the boundaries of a time zone with your watch set to the correct time, at 12 noon the sun would be at its highest point.

Time
zones

As Earth rotates, some of it faces the sun and the rest is in darkness. Since the sun is high in the sky at noon, noon is at different times in different places. We adjust by splitting the Earth into time zones.

Day and night

On the globe of Earth, we can see day and night divided by a straight line from north to south. When the Earth is laid flat as on the map here, the light and dark areas form a bell shape.

Northern summer
The Earth is tilted. When the North Pole tilts toward the sun and the South Pole leans away, it is summer in the northern hemisphere (northern half of the world) and winter in the southern hemisphere, as on the main map.

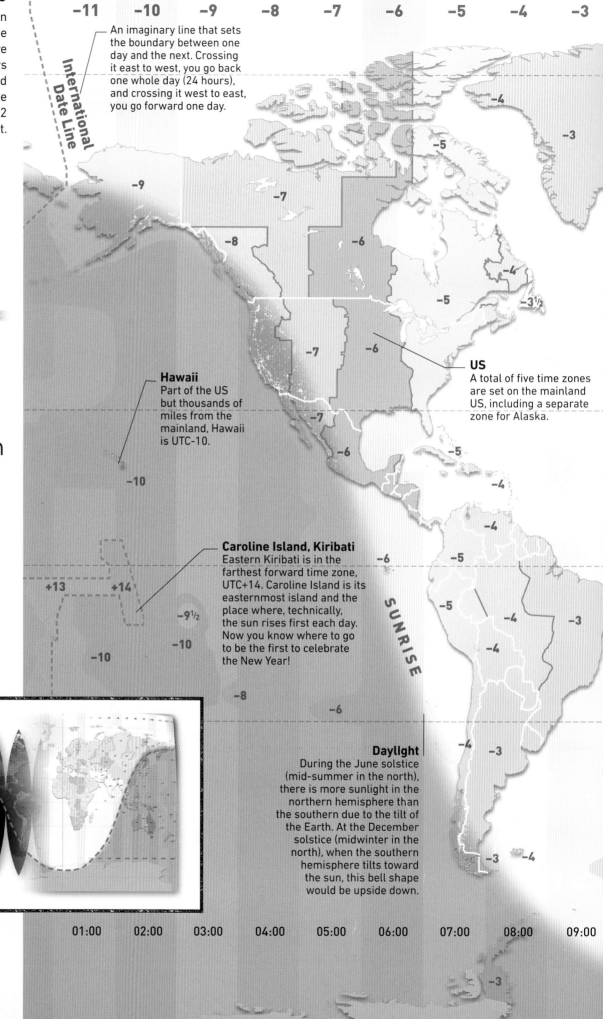

International Date Line

An imaginary line that sets the boundary between one day and the next. Crossing it east to west, you go back one whole day (24 hours), and crossing it west to east, you go forward one day.

Hawaii
Part of the US but thousands of miles from the mainland, Hawaii is UTC-10.

US
A total of five time zones are set on the mainland US, including a separate zone for Alaska.

Caroline Island, Kiribati
Eastern Kiribati is in the farthest forward time zone, UTC+14. Caroline Island is its easternmost island and the place where, technically, the sun rises first each day. Now you know where to go to be the first to celebrate the New Year!

SUNRISE

Daylight
During the June solstice (mid-summer in the north), there is more sunlight in the northern hemisphere than the southern due to the tilt of the Earth. At the December solstice (midwinter in the north), when the southern hemisphere tilts toward the sun, this bell shape would be upside down.

BEFORE TIME ZONES, LOCAL TIME WAS DECIDED BY THE TOWN TIME-

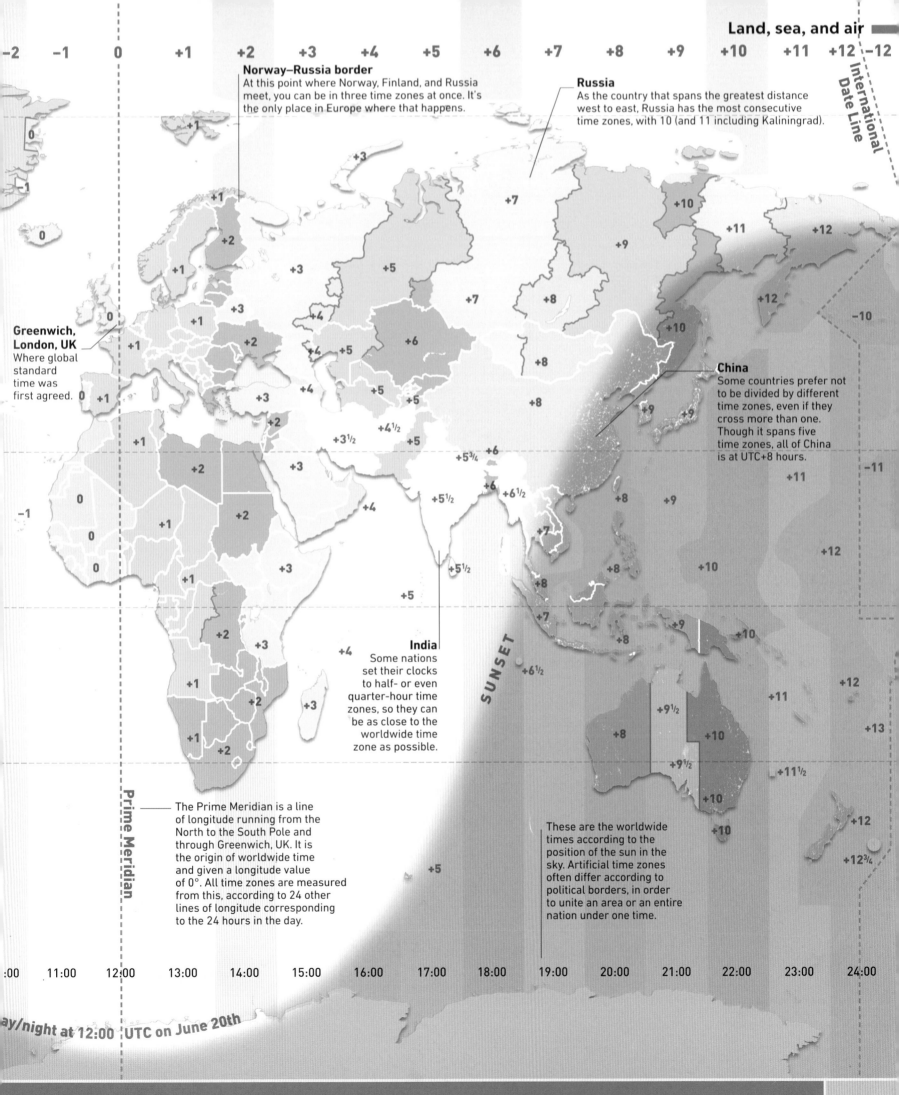

-2   -1   0   +1   +2   +3   +4   +5   +6   +7   +8   +9   +10   +11   +12   -12

International Date Line

Norway–Russia border
At this point where Norway, Finland, and Russia meet, you can be in three time zones at once. It's the only place in Europe where that happens.

Russia
As the country that spans the greatest distance west to east, Russia has the most consecutive time zones, with 10 (and 11 including Kaliningrad).

Greenwich, London, UK
Where global standard time was first agreed.

China
Some countries prefer not to be divided by different time zones, even if they cross more than one. Though it spans five time zones, all of China is at UTC+8 hours.

India
Some nations set their clocks to half- or even quarter-hour time zones, so they can be as close to the worldwide time zone as possible.

SUNSET

Prime Meridian

The Prime Meridian is a line of longitude running from the North to the South Pole and through Greenwich, UK. It is the origin of worldwide time and given a longitude value of 0°. All time zones are measured from this, according to 24 other lines of longitude corresponding to the 24 hours in the day.

These are the worldwide times according to the position of the sun in the sky. Artificial time zones often differ according to political borders, in order to unite an area or an entire nation under one time.

:00   11:00   12:00   13:00   14:00   15:00   16:00   17:00   18:00   19:00   20:00   21:00   22:00   23:00   24:00

ay/night at 12:00 UTC on June 20th

Living world

Humpback whales
Two humpbacks "breach" (leap out of the water) off the coast of Alaska. During winter, humpbacks move south to warmer waters.

Introduction

Life exists in every corner of the planet—from high mountains to deep oceans, and from blazing deserts to the freezing polar regions. Each animal's body, life cycle, and behavior is adapted to its particular habitat, because this maximizes its chances of survival. Plant species, too, have their own adaptations that help them thrive.

Birds

The power of flight allows birds to reach the remotest islands, and some to live in different parts of the world in summer and winter, migrating between the two. There is almost nowhere on Earth that lacks birdlife. Here are their secrets.

- **Lightweight bones**
  Most bird bones are hollow, reinforced by bony struts.

- **Flight feathers**
  Wing and tail feathers provide lift and steer the bird in fight.

- **Warming feathers**
  Two layers of body feathers keep the bird's skin warm.

- **Efficient lungs**
  Bird lungs are far more efficient than mammals', giving them the oxygen they need for energetic flight.

Bald eagle
A North American bird of prey, the bald eagle snatches fish from lakes.

Marine animals

Living in water gives more support than living on land, so many sea creatures survive without strong skeletons. Sea water carries clouds of microscopic life-forms and dead matter, and many sea animals can afford to give up moving from place to place, fix themselves to the seabed, and "filter feed" by grabbing these passing pieces of food.

Coral
Tropical coral reefs are giant growths of filter-feeding life-forms on the seabed.

- **Gills**
  Sea mammals must surface to breathe, but fish take oxygen directly from the water using their gills.

- **Smooth shape**
  Fast-moving marine animals have a streamlined body, which helps them move through the water easily.

- **Buoyancy aid**
  Some fish have an air-filled "swim bladder" to help control buoyancy.

- **Bioluminescence**
  It is dark in the ocean depths. Many deep-sea animals produce light by chemical reactions in their bodies.

Desert cacti
The waxy, fleshy bodies of these desert plants store water. The leaves are reduced to spines, which lose less water to the air. The roots of a cactus may spread out over a wide area, to absorb as much water as possible.

Spineless cactus
A spineless variety of the prickly pear.

THERE ARE PROBABLY AT LEAST 1 MILLION UNDISCOVERED SPECIES IN

Polar regions

The sea in the Arctic and Antarctic is so cold, fish are in danger of freezing. Above the water it is even colder, and no large, cold-blooded animals exist. Warm-blooded animals—those able to retain body heat—predominate. Polar mammals often have two layers of fur: an underlayer of soft hairs that trap air warmed by the animal's body close to the skin, and an outer coat of coarse hairs that keeps out the fiercest gales.

Polar bear
This arctic mammal has a bulky, rounded body surrounded by fat and fur that keep it warm.

- **Natural antifreeze**
  Most polar fish have a chemical in their blood that prevents ice crystals from forming in the body.

- **Small extremities**
  Polar bears and Arctic foxes have small, rounded ears and muzzles that reduce heat loss.

- **Legs and feet**
  Some animals have long legs that wade through snow or broad feet that act like snowshoes.

Western brown snake
A venomous Australian desert species.

Desert regions

The driest parts of the world challenge plants and animals, and desert wildlife is not as abundant as in wetter regions. Desert life-forms must get enough water—and keep what they have. Some desert animals get all the water they need from their food.

- **Nocturnal lifestyle**
  Many animals are active only at night. Gerbils and jerboas retreat into daytime burrows to stay cool.

- **Large extremities**
  Fennec foxes have huge ears that radiate heat away from the body.

- **Drinking dew**
  Insects and lizards drink dewdrops. Larger desert animals that feed at dawn take in dew as they eat plants.

Plant adaptations

In rainforests, plants are in strong competition to reach sunlight. They all grow as fast as possible whenever there is an opening allowing in the sun. In deserts, plants get plenty of light, but they struggle to get enough water from the soil.

Bo tree leaves
This fig tree with drip-tip leaves grows in the rainforests of southern Asia.

Rainforest plants
To reach the sun, many rainforest plants are specialist climbers, and others are epiphytes, which grow on top of other plants. Many rainforest leaves taper to a long point, a "drip tip," to help excess rainwater run off.

THE MUD OF THE DEEP OCEAN FLOOR—PERHAPS UP TO 100 MILLION!

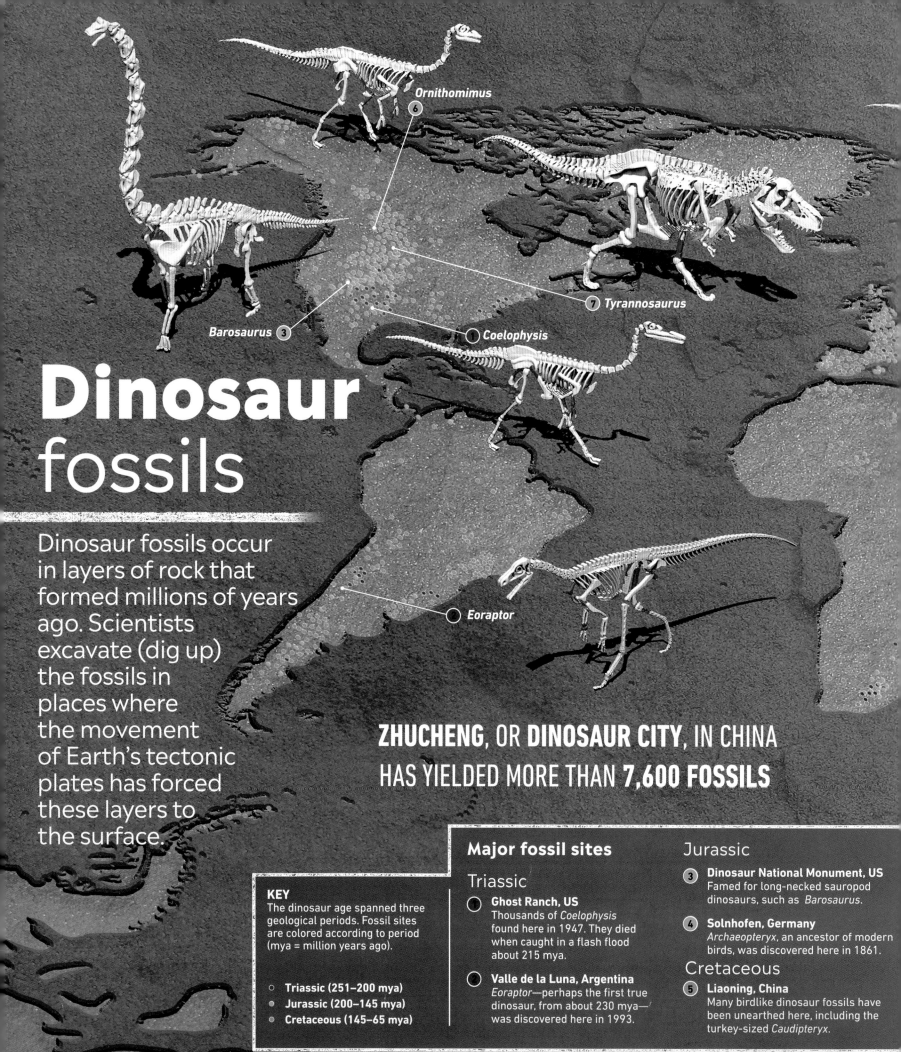

Dinosaur fossils

Dinosaur fossils occur in layers of rock that formed millions of years ago. Scientists excavate (dig up) the fossils in places where the movement of Earth's tectonic plates has forced these layers to the surface.

Ornithomimus (6)

Barosaurus (3)

(7) **Tyrannosaurus**

(1) **Coelophysis**

(2) **Eoraptor**

ZHUCHENG, OR DINOSAUR CITY, IN CHINA HAS YIELDED MORE THAN 7,600 FOSSILS

Major fossil sites

Triassic

(1) **Ghost Ranch, US**
Thousands of *Coelophysis* found here in 1947. They died when caught in a flash flood about 215 mya.

(2) **Valle de la Luna, Argentina**
Eoraptor—perhaps the first true dinosaur, from about 230 mya—was discovered here in 1993.

Jurassic

(3) **Dinosaur National Monument, US**
Famed for long-necked sauropod dinosaurs, such as *Barosaurus*.

(4) **Solnhofen, Germany**
Archaeopteryx, an ancestor of modern birds, was discovered here in 1861.

Cretaceous

(5) **Liaoning, China**
Many birdlike dinosaur fossils have been unearthed here, including the turkey-sized *Caudipteryx*.

KEY
The dinosaur age spanned three geological periods. Fossil sites are colored according to period (mya = million years ago).

- Triassic (251–200 mya)
- Jurassic (200–145 mya)
- Cretaceous (145–65 mya)

AS WELL AS BONES, NESTS, EGGS, AND TRACKS, FOSSILS INCLUDE

Dinosaur footprints

Fossil hunters have found tracks preserved in mud and sand that later turned into rock. These tracks can tell us how dinosaurs walked, and whether they lived alone or in groups. The sites shown here are all in the US.

Dinosaur Ridge
Colorado. Hundreds of prints unearthed when building a road.

Dinosaur State Park
Connecticut. One of the largest track sites in North America.

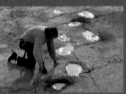

Purgatoire River site
Colorado. Giant sauropod prints left on a lake shore.

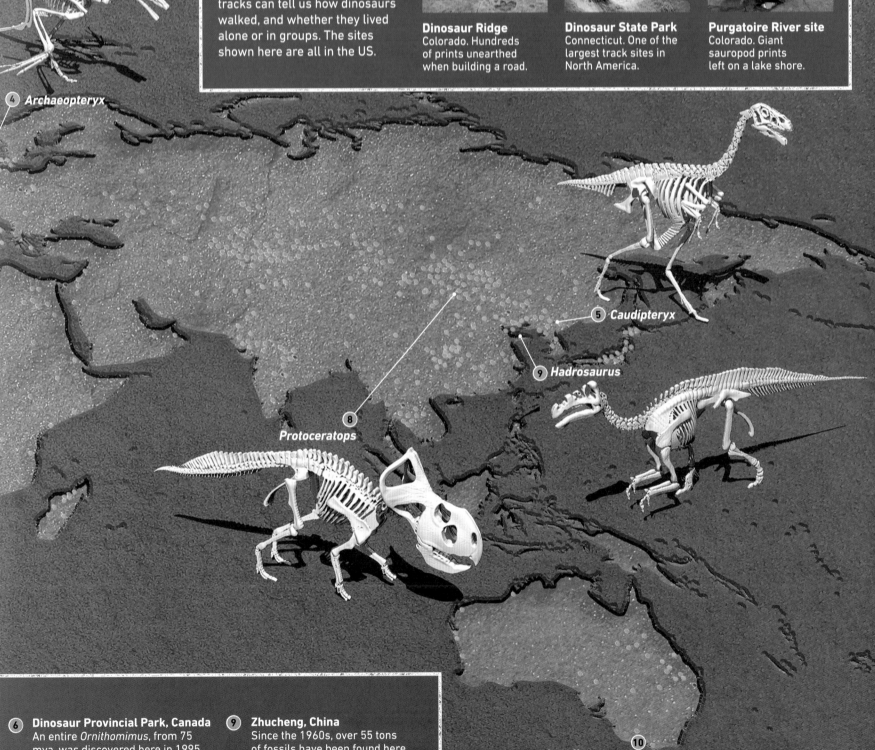

④ *Archaeopteryx*

⑤ *Caudipteryx*

⑨ *Hadrosaurus*

⑧ *Protoceratops*

⑩ *Leaellynasaura*

⑥ **Dinosaur Provincial Park, Canada**
An entire *Ornithomimus*, from 75 mya, was discovered here in 1995.

⑦ **Hell Creek, US**
Ancient rocks here have yielded a range of dinosaur fossils—among them, *Tyrannosaurus*.

⑧ **Flaming Cliffs, Mongolia**
The first *Protoceratops* fossils and dinosaur nest were found here.

⑨ **Zhucheng, China**
Since the 1960s, over 55 tons of fossils have been found here. Rich in remains of "duck-billed" dinosaurs such as *Hadrosaurus*.

⑩ **Dinosaur Cove, Australia**
About 105 mya this was near the South Pole. Until the discovery of *Leaellynasaura* here in 1989, no one knew dinosaurs could live through cold, long, dark winters.

TRACES OF SKIN AND FEATHERS, AND EVEN DINOSAUR POOP!

Americas

1. Bald eagle
Stabs its sharp talons into prey and rips open the body with its hooked bill.

2. Wolverine
Preys on rodents, other small mammals, and even weakened reindeer.

3. Coyote
Eats almost anything, from insects and frogs to calves and lambs.

4. Boa constrictor
A large snake, the boa coils around its prey and squeezes until the victim suffocates.

5. Jaguar
Unable to run fast for very long, the jaguar relies on stealth to creep up on prey.

6. Piranha
Using razor-sharp teeth, a shoal can reduce a deer to bones in minutes.

Africa

7. African rock python
Growing up to 28 ft (8.5 m) long, pythons prey on monkeys, pigs, and birds.

8. African lion
The females do most of the hunting. The male defends the pride's territory.

9. African wild dog
Can chase down prey at 25 mph (40 kph) for 3 miles (5 km) or more.

Eurasia

10. Polar bear
Can kill with a single swipe from one of its 40-lb (18-kg) front paws.

11. Golden eagle
With its amazing eyesight, can spot prey 1.25 miles (2 km) away.

12. Gray wolf
Packs can bring down animals as large as reindeer or musk ox.

---

10. Polar bear
On land and sea ice within the Arctic Circle

11. Golden eagle
Europe, North America, northern Asia, and Africa

1. Bald eagle
Throughout North America

2. Wolverine
Canada and northern US; Scandinavia and Siberia

3. Coyote
From Alaska to Central America

19. Killer whale (orca)
Oceans worldwide

4. Boa constrictor
From Mexico to Argentina

20. Common dolphin
Cool and warm oceans worldwide

3,000–4,000:
THE NUMBER OF **TIGERS** LEFT IN THE WILD

21. Sperm whale
Worldwide, to the edge of the polar ice

7. African rock python
Africa, south of the Sahara

18. California sea lion
Pacific coast of North America and the Galápagos Islands

5. Jaguar
Southwestern US to northern Argentina

Predators

Found on every continent and in every ocean, predators are animals that kill and eat other creatures. With their incredible array of hunting strategies and body parts adapted for killing, they include some of the most fascinating species on the planet.

22. Tuna
Cool and warm oceans worldwide

6. Piranha
North, central, and eastern South America

23. Great white shark
Cool and warm oceans worldwide

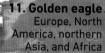

WHEN A PEREGRINE FALCON DIVES ON A PIGEON AT FULL SPEED,

13. Eurasian lynx
Furry ear tufts gather prey noises in the dense forest, where sounds are muffled.

14. Peregrine falcon
Dives onto prey at 200 mph (320 kph), making it the fastest animal on Earth.

15. Eurasian badger
Eats worms, insects, birds, frogs, lizards, and small mammals, plus plants.

16. Tiger
Camouflaged by its stripes, a tiger stalks its prey and kills with a bite to the neck.

17. Sunda clouded leopard
For its size, this shy forest-dweller has longer canine teeth than any other cat.

Oceans

18. California sea lion
May hunt nonstop for 30 hours, diving for up to 5 minutes at a time.

19. Killer whale (orca)
Many hunt sea lions, dolphins, and even whales. Can snatch seals off the ice.

20. Common dolphin
Together, dolphins can herd fish to the surface, where they are easier to catch.

21. Sperm whale
May dive to 9,843 ft (3,000 m) deep in search of giant squid.

22. Tuna
Able to swim at 50 mph (80 kph); hunts fish and squid near surface.

23. Great white shark
Kills dolphins, seals, and big fish, including other sharks, with its jagged teeth.

12. Gray wolf
Much of Asia, parts of Europe, and northern North America

13. Eurasian lynx
Europe (mainly northern and eastern parts) to northern and central Asia

14. Peregrine falcon
Lives on every continent except Antarctica

16. Tiger
Parts of India, China, Siberia, and southeast Asia

15. Eurasian badger
Europe and Asia below the Arctic Circle

8. African lion
Africa, south of the Sahara

17. Sunda clouded leopard
Sumatra and Borneo in southeast Asia

Australasia

24. Saltwater crocodile
Preys on water buffalo and cattle on land. Spends much of its life at sea, catching fish.

25. Tasmanian devil
This marsupial's strong jaws can crush the bones of birds, fish, and small mammals.

9. African wild dog
Africa, south of the Sahara

Food chains

A food chain shows how food energy passes from one living thing to the next. Food chains start with plants, which use sunlight to make their own food. Plants are eaten by herbivores. Predators eat herbivores and smaller predators.

Martial eagle (top predator)

Meerkat (predator)

Imperial scorpion (predator)

Grasshopper (herbivore)

Grass

A FOOD CHAIN IN THE AFRICAN SAVANNA

24. Saltwater crocodile
Southeast Asia and Northern Australia

25. Tasmanian devil
Tasmania, an island off the southeastern tip of Australia

THE FORCE OF THE IMPACT MAY DECAPITATE (BEHEAD) ITS PREY!

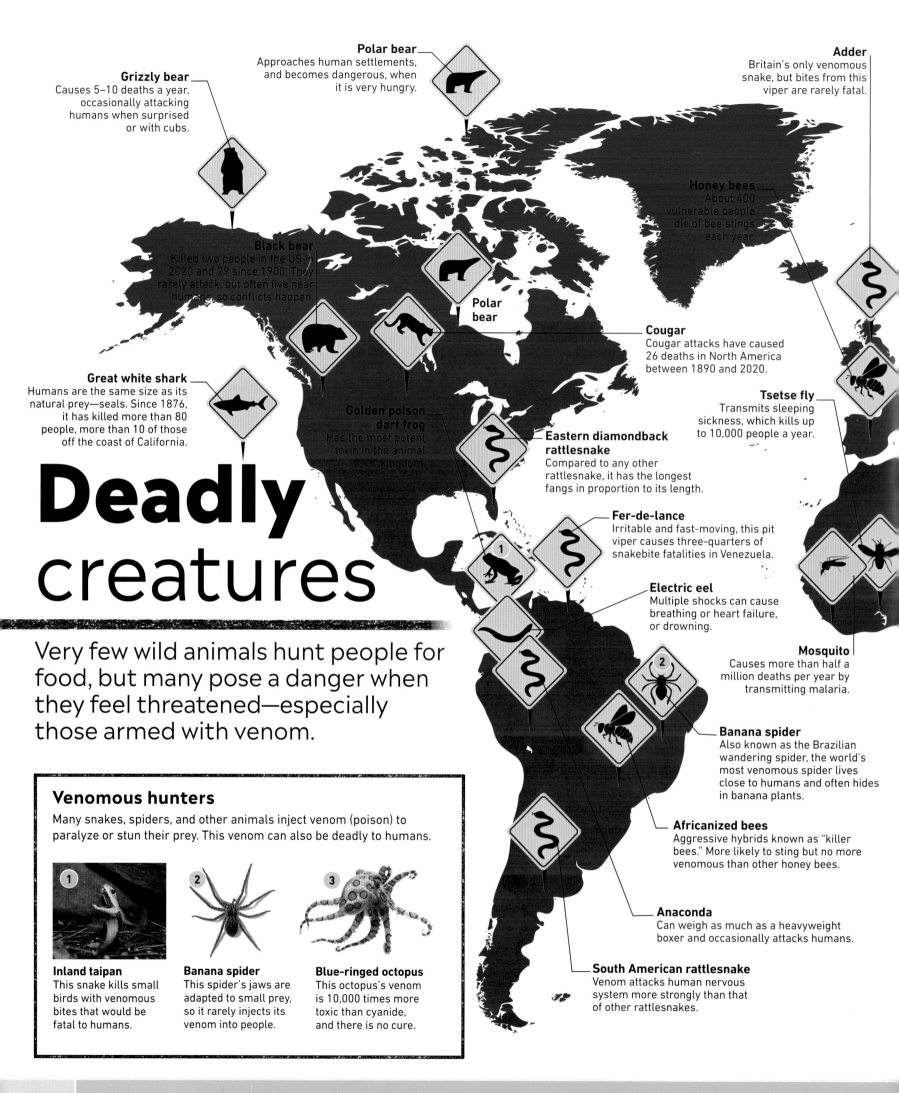

Deadly creatures

Very few wild animals hunt people for food, but many pose a danger when they feel threatened—especially those armed with venom.

Grizzly bear
Causes 5–10 deaths a year, occasionally attacking humans when surprised or with cubs.

Polar bear
Approaches human settlements, and becomes dangerous, when it is very hungry.

Adder
Britain's only venomous snake, but bites from this viper are rarely fatal.

Black bear
Killed two people in the US in 2020 and 29 since 1900. They rarely attack, but often live near humans, so conflicts happen.

Honey bees
About 400 vulnerable people die of bee stings each year.

Polar bear

Cougar
Cougar attacks have caused 26 deaths in North America between 1890 and 2020.

Great white shark
Humans are the same size as its natural prey—seals. Since 1876, it has killed more than 80 people, more than 10 of those off the coast of California.

Golden poison dart frog
Has the most potent toxin in the animal kingdom.

Eastern diamondback rattlesnake
Compared to any other rattlesnake, it has the longest fangs in proportion to its length.

Tsetse fly
Transmits sleeping sickness, which kills up to 10,000 people a year.

Fer-de-lance
Irritable and fast-moving, this pit viper causes three-quarters of snakebite fatalities in Venezuela.

Electric eel
Multiple shocks can cause breathing or heart failure, or drowning.

Mosquito
Causes more than half a million deaths per year by transmitting malaria.

Banana spider
Also known as the Brazilian wandering spider, the world's most venomous spider lives close to humans and often hides in banana plants.

Africanized bees
Aggressive hybrids known as "killer bees." More likely to sting but no more venomous than other honey bees.

Anaconda
Can weigh as much as a heavyweight boxer and occasionally attacks humans.

South American rattlesnake
Venom attacks human nervous system more strongly than that of other rattlesnakes.

Venomous hunters

Many snakes, spiders, and other animals inject venom (poison) to paralyze or stun their prey. This venom can also be deadly to humans.

Inland taipan
This snake kills small birds with venomous bites that would be fatal to humans.

Banana spider
This spider's jaws are adapted to small prey, so it rarely injects its venom into people.

Blue-ringed octopus
This octopus's venom is 10,000 times more toxic than cyanide, and there is no cure.

ANIMAL TOXINS ARE USEFUL. SCIENTISTS HAVE ADAPTED THE

SOME VICTIMS OF **STONEFISH VENOM** SAY IT'S GOOD FOR THEIR ARTHRITIS

Defensive poisons

Many animals use toxins (poisons) against predators. The poisons may be in spines or stings, or they may ooze from the skin.

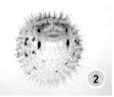

Golden poison dart frog
The skin has enough toxin to kill 10 people. It is effective against its snake predators.

Pufferfish
The poison in puffers' skin and liver could kill a human, but these fish make a prized dish in Japan.

Stonefish
This fish's spines stop predators, but also endanger humans who are pricked by accident.

Asp viper
Causes about 90 percent of all snake bites in Italy, but only 4 percent of bites are fatal.

Pallas's viper
0.004 oz (0.1 gram) of venom can kill a human, but only strikes if threatened.

European black widow spider
Venom is 15 times stronger than a rattlesnake's.

Fat-tailed scorpion
Most dangerous scorpion in North Africa and the Middle East.

Tiger
Until recent improvements in tiger management, hunted and killed about 50 people every year in the Sundarbans mangroves of India.

Pufferfish
Eaten as *fugu* in Japan and *bok-uh* in Korea, but some parts highly poisonous. Accidents happen when untrained people catch and eat the fish.

Common krait
Most venomous land snake in Asia.

Malayan pit viper
Responsible for 700 snakebites annually in Malaysia.

Box jellyfish
Has enough toxin to kill 60 humans, and in the Philippines 20–40 people die each year from stings.

Puff adder
Lives in heavily populated areas and is the most dangerous snake in Africa.

African lion
Kills 70 people a year in Tanzania, either by hunting them for food, or in defense.

Lionfish
Its venomous spines can cause severe injuries, breathing difficulties, and temporary paralysis.

Saltwater crocodile
Makes frequent fatal attacks on humans in New Guinea, the Solomon Islands, and Indonesia.

Elephant
Attacks people when threatened and kills nearly 300 people a year.

Asian cobra
Responsible for more human deaths than any other snake.

Stonefish
Venom injected by spines causes unbearable pain and death in a few hours if not treated.

Hippopotamus
Causes more than 300 deaths a year, sometimes by upturning boats.

Komodo dragon
Giant lizard that grows up to 10 ft (3 m) long and may, very rarely, attack and eat humans.

Cape buffalo
Attacks when defending itself and kills more than 200 people a year.

Blue-ringed octopus
Enough toxin in its body to kill 26 adult humans. It can cause respiratory failure.

Tiger snake
In humans, 60 percent of untreated bites result in serious poisoning or death.

Black mamba
Fastest snake on Earth and kills any human it bites unless the victim takes antivenom.

Redback spider
Also known as the Australian black widow. Deaths are rare, but bites can result in fatal complications.

Six-eyed sand spider
There is no antivenom for its bite but (luckily) it is shy and has little contact with people.

Inland taipan
Deadliest venom of any land snake, but snake scientists are almost the only known victims. They recovered after treatment with antivenom.

Funnel-web spider
Its extremely toxic venom could kill a small child in 15 minutes.

TOXIN FROM POISON DART FROGS TO PRODUCE A POTENT PAINKILLER.

How the aliens invade

Stowaways
Fleas and other parasites can hitch a ride via animal or human hosts. Rats, mice, and insects can travel hidden in ships' cargo. Some species sneak in when empty cargo ships take on local seawater as ballast, then pump it out at their destination. Every day, large numbers of marine organisms are transported around the globe in this way.

Black rat

Introduced by humans
Some species are deliberately introduced by humans. This can be by hunters, for meat, fur, or sport; by farmers; or for biological control, where a new species is introduced to control native pests. Some invaders are escaped pets, or plants washed out of home aquariums. A few have even been released by immigrants who introduce familiar wildlife to remind them of home!

Cane toad

Racoon
Since its introduction, has devastated the seabird population of Canada's Scott Islands.

Zebra mussel
Traveled from the Caspian Sea to the Great Lakes of North America in the ballast water of ships.

Common starling
European native bird released in New York City in 1890 by homesick English settler Eugene Schieffelin.

Stoat
Introduced to islands off Denmark and the Netherlands, it eradicated the native water voles.

Gray squirrel
This US import to Britain competes for habitat with the native red squirrel.

Japanese knotweed
Dense thickets of this weed crowd out native plant life on riverbanks and roadsides in Europe.

Rainbow trout
In California, this fish has endangered the Sierra Nevada yellow-legged frog.

Gypsy moth
This European native costs about $870 million each year in damage to US trees.

Chinese mitten crab
A burrowing species that threatens the US fishery industry by eating bait and trapped fish.

Velvet tree
Known as the "purple plague of Hawaii," it threatens native rainforest plant species.

Flowerpot snake
Emigrated to the US from Africa and Asia by stowing away in the soil of exported pot plants.

Feral pig
In Mexico's Revillagigedo Islands, this former farm animal preys on the endangered Townsend's shearwater bird.

American bullfrog
Native to North America, it is now a resident of more than 40 countries.

Red-vented bulbul
A major agricultural pest in Tahiti, it feeds on fruit and vegetable crops.

Fire ant
Threatens tortoises on the Galápagos Islands by eating hatchlings and attacking adults.

Feral goat
Has caused serious damage to native vegetation on the Galápagos Islands.

Africanized honey bee
Specially bred for survival in the tropics, this "killer bee" turned out to be too aggressive and unpredictable for beekeepers.

ABOUT
90 PERCENT
OF THE WORLD'S **ISLANDS**
HAVE NOW BEEN
INVADED BY RATS

Red Deer
Introduced from Europe to provide sportspeople with game.

House mouse
With no predators on Gough Island, non-native mice have grown to three times their usual size.

INVASIVE SPECIES HAVE PLAYED A PART IN ALMOST HALF OF THE

Alien invasion

Invasive species are animals or plants that enter and thrive in an environment where they are not native. Native species (plants and animals already living there) usually have no defense. The invading aliens can wipe out native species by preying on them or out-competing them.

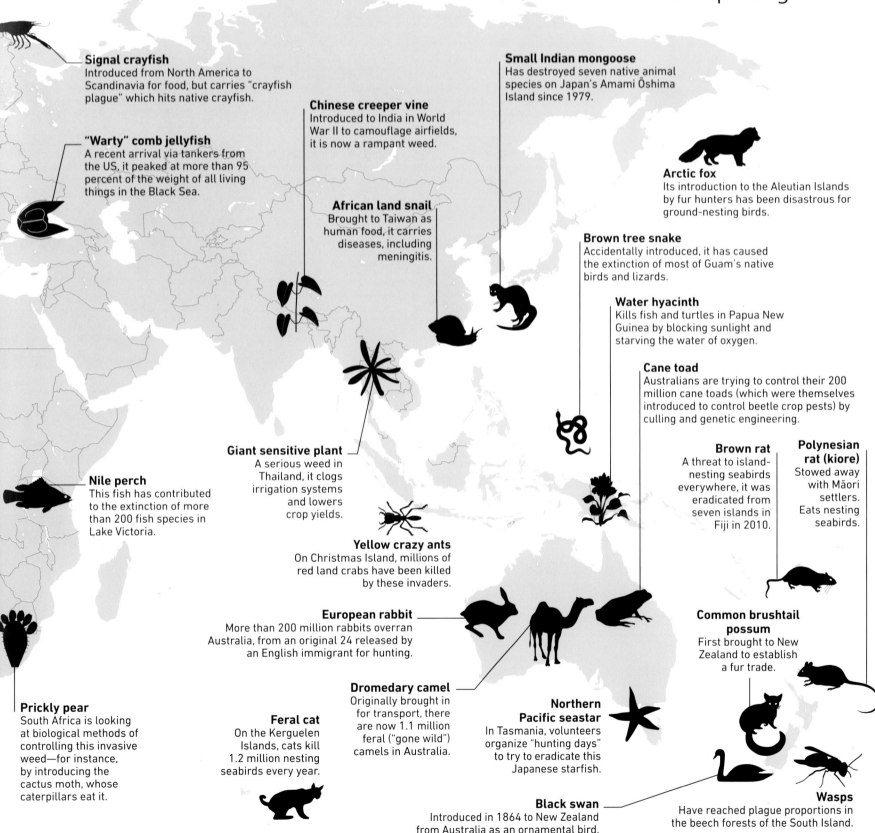

Signal crayfish
Introduced from North America to Scandinavia for food, but carries "crayfish plague" which hits native crayfish.

"Warty" comb jellyfish
A recent arrival via tankers from the US, it peaked at more than 95 percent of the weight of all living things in the Black Sea.

Chinese creeper vine
Introduced to India in World War II to camouflage airfields, it is now a rampant weed.

African land snail
Brought to Taiwan as human food, it carries diseases, including meningitis.

Small Indian mongoose
Has destroyed seven native animal species on Japan's Amami Ōshima Island since 1979.

Arctic fox
Its introduction to the Aleutian Islands by fur hunters has been disastrous for ground-nesting birds.

Brown tree snake
Accidentally introduced, it has caused the extinction of most of Guam's native birds and lizards.

Water hyacinth
Kills fish and turtles in Papua New Guinea by blocking sunlight and starving the water of oxygen.

Cane toad
Australians are trying to control their 200 million cane toads (which were themselves introduced to control beetle crop pests) by culling and genetic engineering.

Nile perch
This fish has contributed to the extinction of more than 200 fish species in Lake Victoria.

Giant sensitive plant
A serious weed in Thailand, it clogs irrigation systems and lowers crop yields.

Yellow crazy ants
On Christmas Island, millions of red land crabs have been killed by these invaders.

European rabbit
More than 200 million rabbits overran Australia, from an original 24 released by an English immigrant for hunting.

Brown rat
A threat to island-nesting seabirds everywhere, it was eradicated from seven islands in Fiji in 2010.

Polynesian rat (kiore)
Stowed away with Māori settlers. Eats nesting seabirds.

Common brushtail possum
First brought to New Zealand to establish a fur trade.

Prickly pear
South Africa is looking at biological methods of controlling this invasive weed—for instance, by introducing the cactus moth, whose caterpillars eat it.

Feral cat
On the Kerguelen Islands, cats kill 1.2 million nesting seabirds every year.

Dromedary camel
Originally brought in for transport, there are now 1.1 million feral ("gone wild") camels in Australia.

Northern Pacific seastar
In Tasmania, volunteers organize "hunting days" to try to eradicate this Japanese starfish.

Black swan
Introduced in 1864 to New Zealand from Australia as an ornamental bird.

Wasps
Have reached plague proportions in the beech forests of the South Island.

ANIMAL EXTINCTIONS THAT HAVE OCCURRED IN THE LAST 400 YEARS.

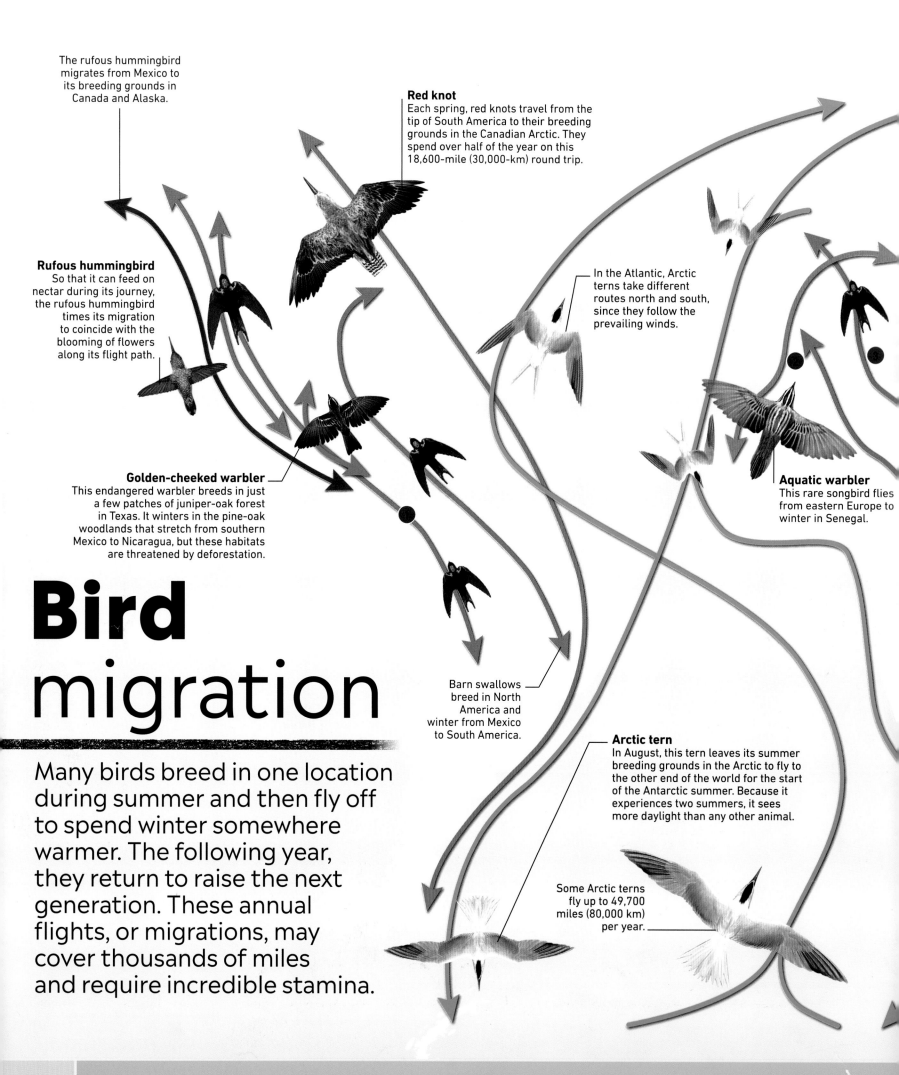

The rufous hummingbird migrates from Mexico to its breeding grounds in Canada and Alaska.

Red knot
Each spring, red knots travel from the tip of South America to their breeding grounds in the Canadian Arctic. They spend over half of the year on this 18,600-mile (30,000-km) round trip.

Rufous hummingbird
So that it can feed on nectar during its journey, the rufous hummingbird times its migration to coincide with the blooming of flowers along its flight path.

In the Atlantic, Arctic terns take different routes north and south, since they follow the prevailing winds.

Golden-cheeked warbler
This endangered warbler breeds in just a few patches of juniper-oak forest in Texas. It winters in the pine-oak woodlands that stretch from southern Mexico to Nicaragua, but these habitats are threatened by deforestation.

Aquatic warbler
This rare songbird flies from eastern Europe to winter in Senegal.

Bird migration

Many birds breed in one location during summer and then fly off to spend winter somewhere warmer. The following year, they return to raise the next generation. These annual flights, or migrations, may cover thousands of miles and require incredible stamina.

Barn swallows breed in North America and winter from Mexico to South America.

Arctic tern
In August, this tern leaves its summer breeding grounds in the Arctic to fly to the other end of the world for the start of the Antarctic summer. Because it experiences two summers, it sees more daylight than any other animal.

Some Arctic terns fly up to 49,700 miles (80,000 km) per year.

Red-breasted goose
After wintering on the Black Sea coast, the red-breasted goose heads north to raise chicks on the Russian tundra.

Barn swallows that spend winter in India fly north to nest in northern Asia.

Ferruginous duck
This widespread duck breeds on marshes and lakes and makes relatively short migrations. Ferruginous ducks that breed in western China and Mongolia winter in India and Pakistan.

A bar-tailed godwit may travel up to 286,000 miles (460,000 km) during the course of its life.

Barn swallows of southern Africa fly to Europe to breed.

Sociable lapwing
In 2007, the sociable lapwing's migration route from east Africa to Kazakhstan and Russia was revealed for the first time by satellite tracking.

Barn swallow
Each year, huge flocks migrate between northern Australia and eastern Russia. These birds can catch insects on the wing and drink by scooping water from lakes.

ARCTIC TERNS FLY FROM THE **ANTARCTIC** TO **GREENLAND** IN 40 DAYS

Aided by strong tailwinds at high altitude, the godwits can make the return journey to New Zealand in just over eight days.

Bar-tailed godwit
Bar-tailed godwits fly from New Zealand to breed in Alaska. On the return trip, one was tracked flying 7,258 miles (11,680 km) nonstop over the Pacific Ocean—the longest continuous journey ever recorded for a bird.

Migration bottlenecks
Places that lie on the flight paths of many birds are known as migration bottlenecks. They are especially important for soaring birds such as storks and birds of prey. These birds can't fly far over water, so they rely on routes with the shortest sea crossings. Millions of birds may pass at these favorite spots.

1. **Panama**
About 3 million birds of prey use this land bridge between North and South America.

2. **Strait of Gibraltar**
Soaring birds fly to Europe from Africa on this sea crossing of only 9 miles (14 km).

3. **Sicily and Malta**
These islands are "stepping stones" for birds flying from Italy to Tunisia and Libya.

4. **Egypt**
Egypt has several bottlenecks—such as Suez, Hurghada, and Zaranik—for birds flying between Africa and Europe or Asia.

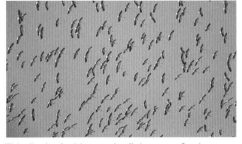

This flock of white storks flying over Spain reached Europe via the Strait of Gibraltar.

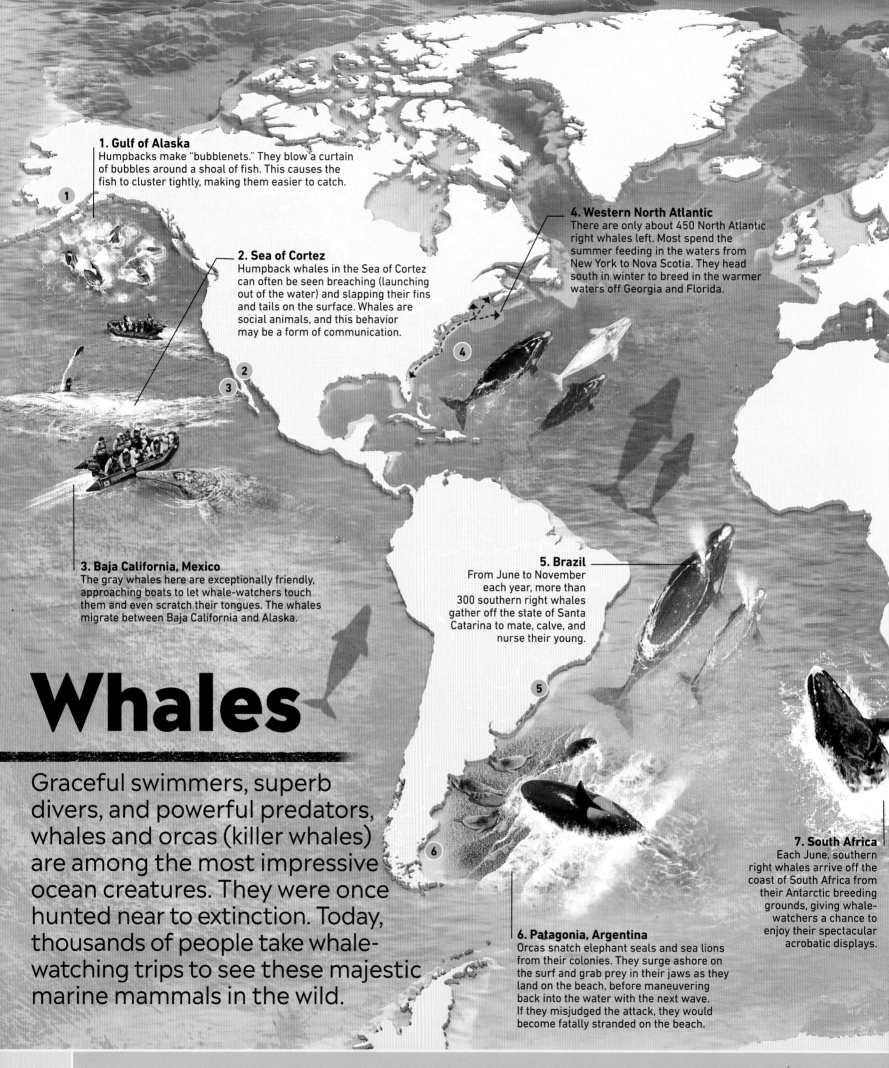

1. Gulf of Alaska
Humpbacks make "bubblenets." They blow a curtain of bubbles around a shoal of fish. This causes the fish to cluster tightly, making them easier to catch.

2. Sea of Cortez
Humpback whales in the Sea of Cortez can often be seen breaching (launching out of the water) and slapping their fins and tails on the surface. Whales are social animals, and this behavior may be a form of communication.

3. Baja California, Mexico
The gray whales here are exceptionally friendly, approaching boats to let whale-watchers touch them and even scratch their tongues. The whales migrate between Baja California and Alaska.

4. Western North Atlantic
There are only about 450 North Atlantic right whales left. Most spend the summer feeding in the waters from New York to Nova Scotia. They head south in winter to breed in the warmer waters off Georgia and Florida.

5. Brazil
From June to November each year, more than 300 southern right whales gather off the state of Santa Catarina to mate, calve, and nurse their young.

6. Patagonia, Argentina
Orcas snatch elephant seals and sea lions from their colonies. They surge ashore on the surf and grab prey in their jaws as they land on the beach, before maneuvering back into the water with the next wave. If they misjudged the attack, they would become fatally stranded on the beach.

7. South Africa
Each June, southern right whales arrive off the coast of South Africa from their Antarctic breeding grounds, giving whale-watchers a chance to enjoy their spectacular acrobatic displays.

Whales

Graceful swimmers, superb divers, and powerful predators, whales and orcas (killer whales) are among the most impressive ocean creatures. They were once hunted near to extinction. Today, thousands of people take whale-watching trips to see these majestic marine mammals in the wild.

THE BLUE WHALE IS THE LARGEST ANIMAL EVER TO HAVE LIVED

1 MILLION SPERM WHALES WERE KILLED BEFORE HUNTING THEM WAS BANNED IN 1981

Migration

Whales travel to cold waters near the poles to feed, then move to warmer waters closer to the equator to breed. Few species migrate across the equator, so there can be separate populations in the northern and southern hemispheres.

KEY

- **Breeding areas** Warmer waters for giving birth
- **Feeding areas** Cooler waters that are rich in food
- ←- -→ **Migration routes** Breeding-to-feeding areas and back
- **Site of spectacular whale behavior**

10. Northwest Pacific
In winter, the humpbacks of the western Pacific mate and calve in warm, subtropical waters from the Philippines to Japan. Summer sees them traveling to feed in the extreme north of the Pacific, around the Aleutian Islands.

8. Sri Lanka
Between December and April, Dondra Point, on Sri Lanka's southern tip, is the best place to see blue whales. Unlike most populations of blue whales, this one does not migrate to polar waters to feed. These northern Indian Ocean blue whales both breed and feed year round in tropical waters.

11. Kaikoura, New Zealand
This one of few places in the world where sperm whales can be seen year round. They are attracted by an underwater canyon close to the shore that has abundant marine life, including the giant squid that the whales hunt.

9. Antarctica
Antarctic orcas often hunt in teams, herding their prey together before attacking from different angles. They will also tip over ice floes to knock penguins and seals into the water.

ON EARTH. ITS TONGUE ALONE CAN WEIGH AS MUCH AS AN ELEPHANT!

SOME **SHARKS** GROW UP TO **30,000 TEETH** IN THEIR **LIFETIME**

Freshwater sharks

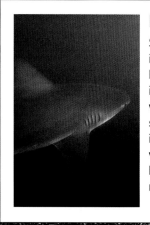

Some shark species are found in freshwater habitats. The bull shark, for example, lives in warm coastal waters worldwide, but it sometimes swims up larger rivers and into lakes. Bull sharks are very territorial, so if they find humans swimming in their river, they may attack them.

Mississippi River
One bull shark reached Alton, Illinois, 1,150 miles (1,850 km upstream.

Potomac River
Bull sharks up to 8 ft (2.4 m) long have been caught in the Potomac.

Lake Nicaragua
Bull sharks reach the lake via the San Juan River.

Amazon River
There have been sightings of bull sharks 1,200 miles (2,000 km) from the sea.

Nicole

In 2003–04, a female great white shark, nicknamed Nicole, made the longest known migration by a shark. Nicole swam from Africa to Australia and back—more than 12,400 miles (20,000 km)—in 9 months. She mostly swam at the surface, but at times she reached depths of up to 3,200 ft (980 m).

- - - - - - - - ->

Nicole's route was tracked using an electronic tag fitted to her fin.

DISTRIBUTION OF SHARKS WORLDWIDE
Some shark species cruise almost all the world's oceans, while others have a more limited range, preferring either cooler or warmer seas.

Whale shark
The largest fish in the sea, reaching lengths of 40 ft (12 m) or more, the whale shark prefers warm waters. It feeds mainly on plankton.

Basking shark
At 30 ft (10 m) long, this is the second-largest fish. Found in temperate seas, it swims open-mouthed, filtering plankton from the sea.

Great white shark
Found in the majority of the world's seas, the great white has made the most recorded attacks on humans. It can swim at more than 25 mph (40 kph).

Great hammerhead shark
Often found near tropical reefs, the great hammerhead preys on stingrays, using its hammer to pin down the fish before biting them.

Port Jackson shark
A reef-dweller from around southern Australia, this shark has wide, flat teeth that crush hard-shelled prey such as oysters, snails, and crabs.

Pygmy shark
At 8–10 in (20–25 cm) long, this is one of the smallest sharks. It hunts squid at depths of up to 6,000 ft (1,800 m) in subtropical and temperate seas.

A GREAT WHITE SHARK'S SUPERB SENSE OF SMELL CAN DETECT TINY

Sharks

Fast, powerful, and armed with razor-sharp teeth, sharks are superb predators. They are much feared, but attacks on people are relatively rare. Humans, in contrast, kill 100 million sharks per year.

Subarctic species
Piked dogfish inhabit temperate and cool seas, venturing as far north as the edge of the Arctic Circle.

Ganges River
In the Ganges and Brahmaputra, the bull shark is often mistaken for the rare Ganges shark.

Zambezi River
Bull sharks are known to attack young hippos.

Wide distribution
The great white shark has one of the greatest ranges of any shark species. However, it is not found in polar waters.

Nicole's route
The trip from South Africa to Australia took Nicole the great white shark 99 days. After about 3 months, she set off again on the return journey.

Pacific angel shark
This shark of the eastern Pacific lies on the seabed and ambushes passing fish. It is superbly camouflaged by its mottled, sandy back.

Ornate wobbegong
Elaborately patterned and with fleshy projections around its jaws, this shark inhabits tropical waters, mainly around the Australian coast.

Frilled shark
With its flat head and eellike body, this frilled shark looks very different than other sharks. It lives near the seabed in deep water.

Longnose sawshark
The longnose lives off southern Australia. Its snout is a long, sawlike projection edged with rows of large, sharp teeth.

Bull shark
This shark is one of the most dangerous to humans. It preys on sharks, rays, and other fish, as well as squid, turtles, and crustaceans.

Piked dogfish
Once among the most abundant sharks, the piked dogfish is now threatened as a result of overfishing. It gathers in shoals by the thousand.

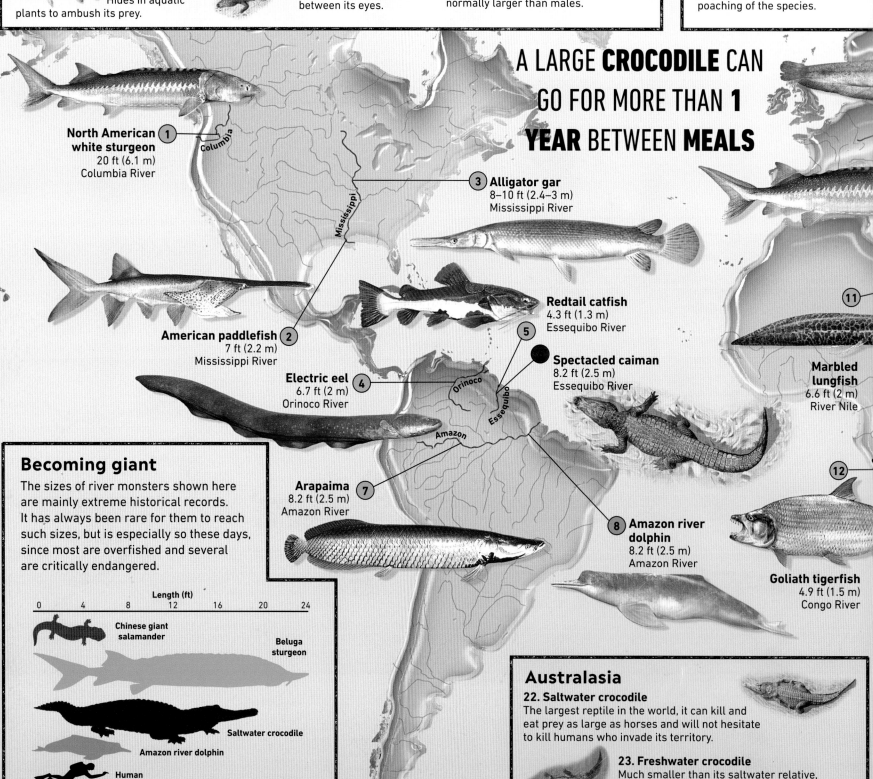

Americas

1. North American white sturgeon
Similar to sturgeons living 100 million years ago, this fish depends heavily on its sense of smell.

2. American paddlefish
Takes its name from its long, paddle-shaped snout.

3. Alligator gar
Hides in aquatic plants to ambush its prey.

4. Electric eel
Generates huge electric shocks to stun prey and ward off attackers.

5. Redtail catfish
Stops feeding to shed its skin like a snake.

6. Spectacled caiman
Named after the bony ridge between its eyes.

7. Arapaima
The adult fish relies on air-breathing, not gills, to get oxygen. But its need to come to the surface makes it vulnerable to hunters.

8. Amazon river dolphin
Hunts in the murky water by sonar and uses its long snout to catch prey hiding in underwater plants. Females are normally larger than males.

Eurasia

9. Wels catfish
Uses its fins to capture prey before swallowing its catch whole.

10. Beluga sturgeon
The world's largest river fish, it spends some of its life in salt water. Extra-large beluga no longer exist due to persistent overfishing and poaching of the species.

A LARGE CROCODILE CAN GO FOR MORE THAN 1 YEAR BETWEEN MEALS

North American white sturgeon 1
20 ft (6.1 m)
Columbia River

3 Alligator gar
8–10 ft (2.4–3 m)
Mississippi River

Redtail catfish
4.3 ft (1.3 m)
Essequibo River

American paddlefish 2
7 ft (2.2 m)
Mississippi River

Electric eel 4
6.7 ft (2 m)
Orinoco River

5 **Spectacled caiman**
8.2 ft (2.5 m)
Essequibo River

Marbled lungfish
6.6 ft (2 m)
River Nile

Arapaima 7
8.2 ft (2.5 m)
Amazon River

8 **Amazon river dolphin**
8.2 ft (2.5 m)
Amazon River

Goliath tigerfish
4.9 ft (1.5 m)
Congo River

Becoming giant

The sizes of river monsters shown here are mainly extreme historical records. It has always been rare for them to reach such sizes, but is especially so these days, since most are overfished and several are critically endangered.

Length (ft)
0 4 8 12 16 20 24

Chinese giant salamander

Beluga sturgeon

Saltwater crocodile

Amazon river dolphin

Human

0 1 2 3 4 5 6 7
Length (m)

Australasia

22. Saltwater crocodile
The largest reptile in the world, it can kill and eat prey as large as horses and will not hesitate to kill humans who invade its territory.

23. Freshwater crocodile
Much smaller than its saltwater relative, it will not attack humans unless provoked.

58   IN ANCIENT JAPANESE FOLKLORE, A GIANT CATFISH, NAMAZU,

Africa

11. Marbled lungfish
In the dry season, digs itself into a mud cocoon for up to 2 years.

12. Goliath tigerfish
Fierce fish known to attack humans.

13. Nile perch
When brought to live in new rivers and lakes, it can kill so many fish that it causes the extinction of native fish species.

Asia

14. Giant devil catfish
This rare species has sharp teeth similar to a shark's.

15. Wallago
Human remains have been found inside its stomach.

16. Gavial
An endangered crocodilian with a long, thin snout, good for catching fish. Rarely grows to 23 ft (7 m).

17. Chinese giant salamander
The world's largest living amphibian.

18. Giant freshwater stingray
Finds its prey using an electric field sensor.

19. Kaluga
Cannibalism is common among these sturgeons of the Russian Far East.

20. Taimen
The largest of the salmon family, also called the "Mongolian terror trout."

21. Giant pangasius
Also known as the "dog-eating catfish." Another critically endangered fish.

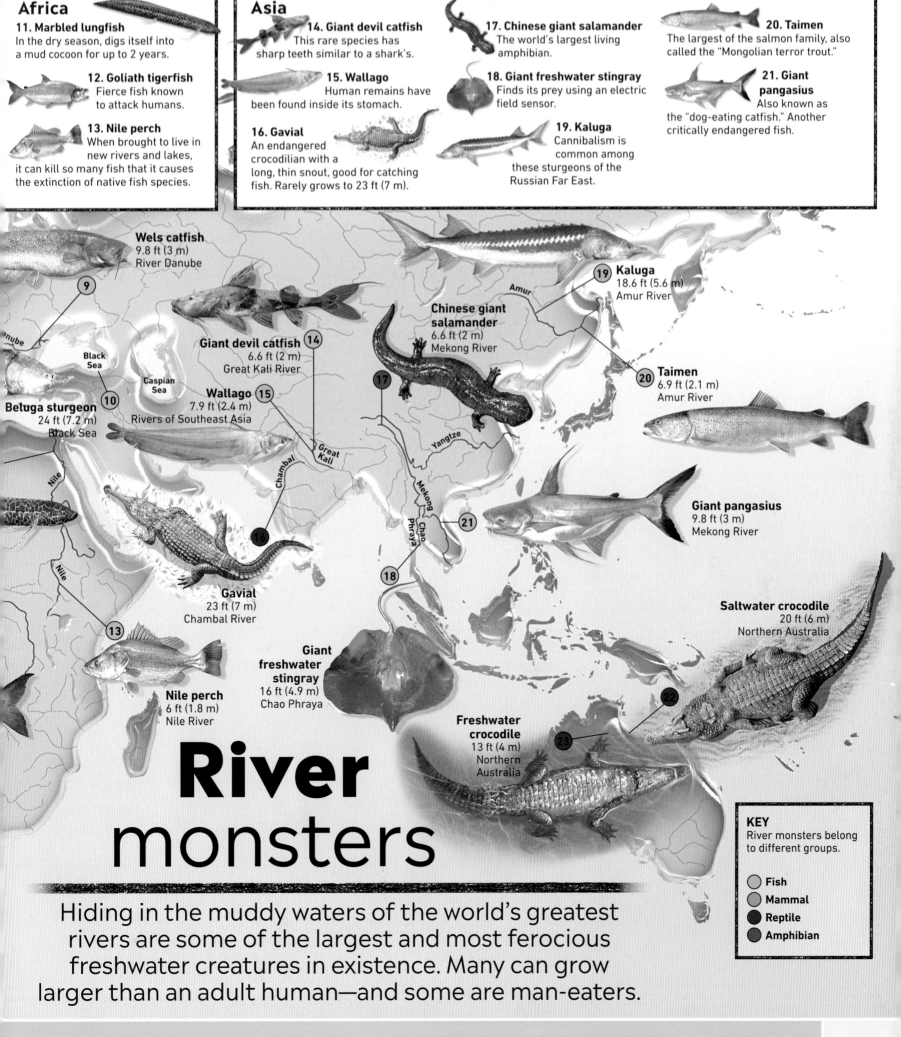

Wels catfish
9.8 ft (3 m)
River Danube
9

Kaluga
18.6 ft (5.6 m)
Amur River
19

Chinese giant salamander
6.6 ft (2 m)
Mekong River

Giant devil catfish 14
6.6 ft (2 m)
Great Kali River

Taimen 20
6.9 ft (2.1 m)
Amur River

Beluga sturgeon
24 ft (7.2 m)
Black Sea
10

Wallago 15
7.9 ft (2.4 m)
Rivers of Southeast Asia

17

Giant pangasius
9.8 ft (3 m)
Mekong River

16

Gavial
23 ft (7 m)
Chambal River

18

Saltwater crocodile
20 ft (6 m)
Northern Australia

13

Nile perch
6 ft (1.8 m)
Nile River

Giant freshwater stingray
16 ft (4.9 m)
Chao Phraya

21

Freshwater crocodile
13 ft (4 m)
Northern Australia
23

22

Black Sea
Caspian Sea
Chambal
Great Kali
Yangtze
Mekong
Chao Phraya
Amur
Nile
Danube

River
monsters

Hiding in the muddy waters of the world's greatest rivers are some of the largest and most ferocious freshwater creatures in existence. Many can grow larger than an adult human—and some are man-eaters.

KEY
River monsters belong to different groups.

- Fish
- Mammal
- Reptile
- Amphibian

KEY
Some insects are found the world over, and some survive only in specific habitats and locations. The insects shown in this map are in locations where they are frequently found.

Insect swarms

Insect record-breakers

Types of swarms

When insects form a large group that moves as a single unit, it is called a swarm. Insects sometimes migrate in swarms, or they swarm when looking for a new home, a mate, or for food.

1

2

5

4

3

6

Maricopa harvester ant
Most venomous.
12 stings can kill a rat.

Mayflies
Shortest adult life.
Mayflies spend most of their lives as water-living nymphs. They transform into winged adults that live just long enough to mate and lay eggs. The most extreme example is the American sand-burrowing mayfly, whose adult life lasts just a few minutes.

Rhyniognatha
Earliest.
A 400-million-year-old fossil was found in Scotland in 1919. Scientists believe it may have been winged.

10

Fairy wasp
Smallest.
0.006 in (0.14 mm) long. Only visible under a powerful microscope.

Termite queen
Longest life.
Can live up to 45 years.

11

Goliath Beetle
Heaviest larva.
Weighs up to 3.5 oz (100 g).

Swarming insects

1 Asian ladybug
Swarm through Oregon in the fall, looking for somewhere to hibernate for the winter.

2 Army cutworm moths
Six- to eight-week migration from eastern plains of Colorado to the mountains.

3 Monarch butterfly
The long migration from the northern US to Mexico lasts generations—no one butterfly makes the entire journey.

4 Termites
In New Orleans, Louisiana, termites build colonies by invading people's homes.

5 Cicadas
In the eastern states of the US, cicada swarms have 13- or 17-year cycles. Young cicadas, known as nymphs, mature, mate, and then die.

6 Mayflies
Annual mass hatching from Lake Erie. They mate, reproduce, then die.

7 Army ants
Found in Central and South America, swarms are called "raids" made up of 100,000–2,000,000 adults.

8 Africanized bees
Aggressive hybrid first released in São Paulo, Brazil. Swarm in thousands when forming new colonies.

9 Dragonflies
A single swarm in Argentina in 1991 was estimated to contain 4–6 billion migrating dragonflies.

10 Flying ants
Swarm annually in Britain as part of a mating ritual.

11 Driver ants
Found in central and east Africa, vast swarms kill animals in their path. People who cannot move out of the way, such as the sick or injured, can be killed.

7

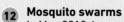

8

9

12 Mosquito swarms
In May 2012, immense swarms of mosquitoes hatched from a lake near Mikoltsy, Belarus.

13 Locusts
The largest swarm recorded was in Kenya in 1954. It covered 77 sq miles (200 sq km) and involved an estimated 10 billion locusts.

14 Midges
The midges that form mating swarms start out as underwater larvae in lakes. Once they can fly, they take off and try to find a mate.

Honey bees
Bees swarm when they leave their hive to find a new home. Once a small number of special "scouts" have agreed on the most suitable site, the queen and the main cluster of bees fly to the new location.

Monarch migration
Every year, by instinct alone, millions of monarch butterflies travel up to 2,500 miles (4,000 km) from northern parts of America to warmer climates as far south as Mexico, before they return north in spring.

Midges
Huge swarms appear over Lake Victoria in Africa during the annual mating season, as thousands of dancing male midges try to attract females. Swarms are so big, they look like giant brown clouds.

Froghopper
Highest jumper.
Jumps 28 in (71 cm)—150 times its own height, which is comparable to a human jumping over a 60-story building!

12

Himalayan cicada
Loudest.
Calls at up to 120 decibels—as loud as an ambulance siren.

Stink bug
Smelliest.
Toxic odor can be smelled by humans about 3.3–5 ft (1–1.5 m) away.

Flea
Longest jumper.
Can jump more than 200 times its body length.

13

14

Dung beetle
Strongest.
Can pull 1,141 times its own body weight—the equivalent to an average human pulling six double-decker buses full of people.

Chan's megastick
Longest.
22.3 in (56.7 cm). Only six specimens have ever been found, all on the island of Borneo.

SCIENTISTS ESTIMATE
4–20 MILLION
TYPES OF **INSECTS** HAVE
YET TO BE **DISCOVERED**

Australian tiger beetle
Fastest runner.
5.6 mph (9 kph). Equivalent to a human running at 480 mph (770 kph).

Giant weta
Heaviest.
Weighs up to 2.5 oz (70 g)—heavier than a sparrow.

Insects

We know of more than 1 million different types of insects, and more are identified every year. They have fascinating habits, and their strange appearances can be seen with the help of microscopes and special cameras.

Horsefly
Fastest flyer.
Maximum speed recorded briefly on takeoff at 90 mph (145 kph). The next fastest are dragonflies and hawk moths, at arbout 30–35 mph (50–55 kph).

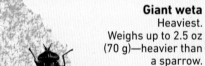

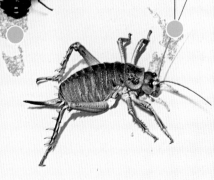

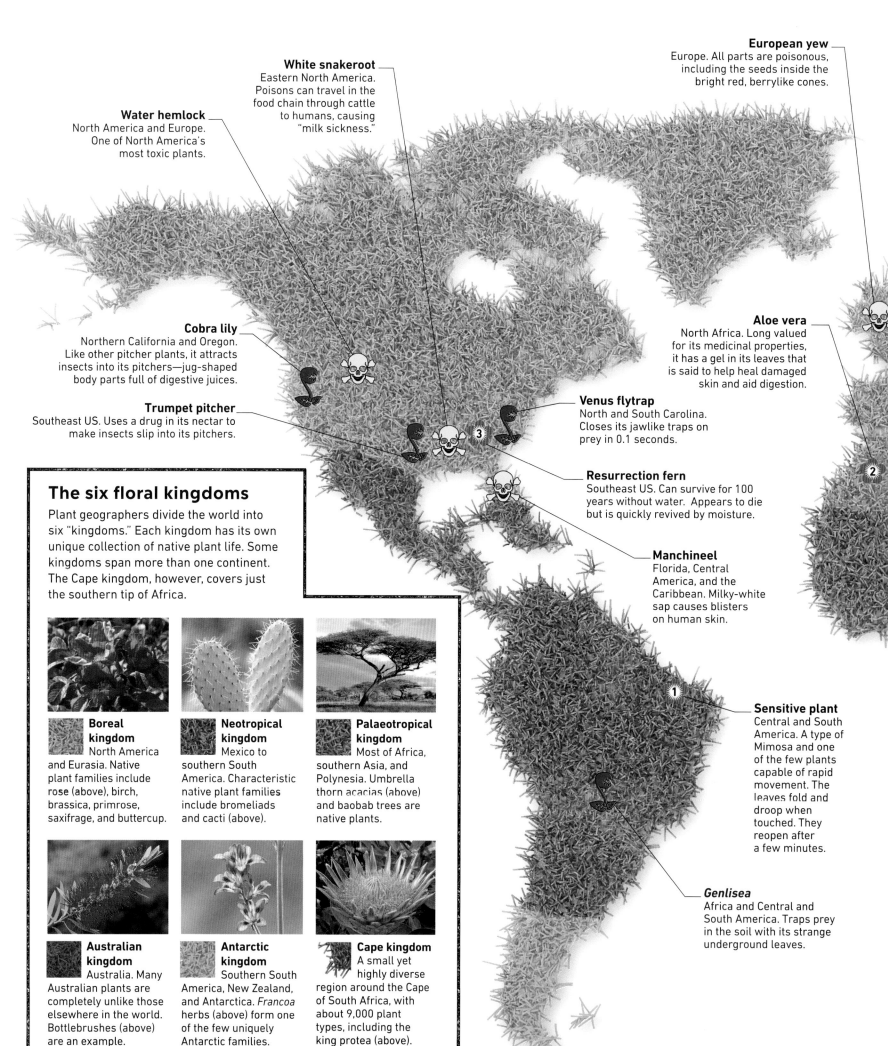

Water hemlock
North America and Europe. One of North America's most toxic plants.

White snakeroot
Eastern North America. Poisons can travel in the food chain through cattle to humans, causing "milk sickness."

European yew
Europe. All parts are poisonous, including the seeds inside the bright red, berrylike cones.

Cobra lily
Northern California and Oregon. Like other pitcher plants, it attracts insects into its pitchers—jug-shaped body parts full of digestive juices.

Trumpet pitcher
Southeast US. Uses a drug in its nectar to make insects slip into its pitchers.

Aloe vera
North Africa. Long valued for its medicinal properties, it has a gel in its leaves that is said to help heal damaged skin and aid digestion.

Venus flytrap
North and South Carolina. Closes its jawlike traps on prey in 0.1 seconds.

Resurrection fern
Southeast US. Can survive for 100 years without water. Appears to die but is quickly revived by moisture.

Manchineel
Florida, Central America, and the Caribbean. Milky-white sap causes blisters on human skin.

Sensitive plant
Central and South America. A type of Mimosa and one of the few plants capable of rapid movement. The leaves fold and droop when touched. They reopen after a few minutes.

Genlisea
Africa and Central and South America. Traps prey in the soil with its strange underground leaves.

The six floral kingdoms

Plant geographers divide the world into six "kingdoms." Each kingdom has its own unique collection of native plant life. Some kingdoms span more than one continent. The Cape kingdom, however, covers just the southern tip of Africa.

Boreal kingdom
North America and Eurasia. Native plant families include rose (above), birch, brassica, primrose, saxifrage, and buttercup.

Neotropical kingdom
Mexico to southern South America. Characteristic native plant families include bromeliads and cacti (above).

Palaeotropical kingdom
Most of Africa, southern Asia, and Polynesia. Umbrella thorn acacias (above) and baobab trees are native plants.

Australian kingdom
Australia. Many Australian plants are completely unlike those elsewhere in the world. Bottlebrushes (above) are an example.

Antarctic kingdom
Southern South America, New Zealand, and Antarctica. *Francoa* herbs (above) form one of the few uniquely Antarctic families.

Cape kingdom
A small yet highly diverse region around the Cape of South Africa, with about 9,000 plant types, including the king protea (above).

Butterwort
Boggy parts of Europe, North and South America, and Asia. Sticky hairs on its leaves trap insects.

Monkshood
Mountains of the northern hemisphere. Also known as aconite, it is a source of a deadly poison contained in the seeds.

KEY

 Poisonous plants
Some plants contain toxic chemicals. The map shows eight of the most poisonous.

 Carnivorous plants
These plants trap and consume insects and other small creatures.

 Incredible plants
Four amazing plants are highlighted on the map, but there are many thousands more worldwide.

Sundew
Worldwide in boggy places. Traps insects with droplets of glue coating its leaves.

Waterwheel plant
Africa, Asia, Australia, and Europe. Freshwater plant a little like an underwater Venus flytrap.

Deadly nightshade
Europe, north Africa, and west Asia

Castor oil plant
East Africa, Mediterranean, and India. Origin of the poison ricin.

Nepenthes rajah
Borneo. This giant pitcher plant may sometimes catch rats or lizards to eat.

Rosary pea
Indonesia. Toxins are used in herbal medicines of southern India.

Welwitschia
Namib Desert. Has just two straplike leaves. They can grow up to 20 ft (6.2 m) long over several centuries.

World of plants

Scientists estimate there are at least 400,000 species of plants on Earth— and possibly many thousands more. Some parts of the world have a rich diversity of plant life; in others, such as Antarctica, plants are scarce.

Rainbow plant
Australia. Catches insects on its sticky leaves.

 Terrestrial bladderwort
Worldwide. Grows on wet, rocky surfaces and catches tiny prey in bladderlike traps.

Total number of life-forms

There are many thousands of species of vertebrate animals, such as birds and reptiles. But these numbers are dwarfed by the amazing number of other life-forms, particularly insects.

NUMBER OF KNOWN SPECIES IN EACH GROUP	
13,000	Algae
74,000	Fungi
17,000	Lichens
320,000	Plants
85,000	Mollusks (squid, clams, snails, and relatives)
47,000	Crustaceans (crabs, shrimps, and relatives)
102,000	Arachnids (spiders, scorpions, and relatives)
1,000,000	Insects
71,000	Other invertebrates (without backbones)
62,000	Vertebrates (animals with backbones)

70,000 weevils

Weevils form only one family of beetles, yet there are more different types than all the world's vertebrates.

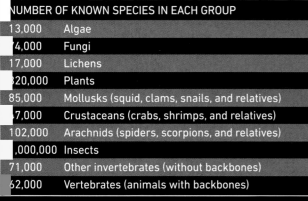

Giraffe-necked weevil

Cratosomus roddami, a weevil

Eupholus linnei, a weevil

Biodiversity

Richness of different life-forms, or species, is called biodiversity. Places such as tropical rainforests are naturally high in biodiversity. Harsh environments have fewer species, but those species might be unique and equally precious.

Barren Arctic
Plants grow very slowly in the cold Canadian Arctic, so there is not a lot of food to go round. Vegetation is ground-hugging, with little variety of homes for small animals—unlike forests. Biodiversity is low.

Rich Amazon
The Amazon is the largest and most diverse tropical forest on Earth. In general, large, continuous areas of habitat support the greatest diversity of species.

Deserted Sahara
There are hardly any amphibians in this dry environment, but the few that survive here are uniquely adapted to the conditions. Preserving areas of pristine Sahara would ensure the survival of some rare creatures.

Unique Atlantic Forest
What remains of the rainforest region in Brazil is not only rich in species. Because it is isolated from other rainforests, many of its species are also found nowhere else.

KEY
This map shows the pattern of biodiversity across the world's land, combining measures of 5,700 mammal species, 7,000 amphibians, and 10,000 species of birds. This gives an overall measure, because the variety of these three groups usually mirrors the total biodiversity, including the numbers of different insects and plants. Scientists know biodiversity in the oceans is lower than on land, but it is not shown on the map.

Lowest   Highest

BIODIVERSITY (SPECIES RICHNESS)

SCIENTISTS ESTIMATE THAT GLOBAL BIODIVERSITY HAS FALLEN TO 84.6

A few tough species
Only a few animal species have what it takes to survive in cold habitats such as the Russian Arctic.

Diverse tropical Asian forests
Tropical rainforest is the most biodiverse habitat. It has abundant water and no shortage of food. The trees provide a multitude of animal homes, from their roots up to their crowns. The climate changes little. All these things allow plants and animals to diversify by evolution into thousands of species.

Borneo
Scientists found an amazing 1,200 tree species here within a tiny plot of rainforest.

Himalayas and Hundu Kush
This mountainous region is home to 25,000 plant species, or nearly 10 percent of the world's total.

Varied African highlands
Mountains are diverse places because they contain a range of different conditions at different heights. At each height lives a different community of plants and animals adapted to those conditions.

LIFE ON LAND IS AS MUCH AS 25 TIMES AS VARIED AS LIFE IN THE SEA

POISON-DART FROGS
There are 175 species in the poison-dart frog family, which lives in the tropical rainforests of Central and South America. They are all related, but each has evolved slightly differently.

Mimic poison-dart frog

Granular poison-dart frog

Three-striped poison-dart frog

Yellow-banded poison-dart frog

Brazil-nut poison-dart frog

Golden poison-dart frog

PERCENT OF ITS LEVEL BEFORE PEOPLE CHANGED THE LANDSCAPE.

Unique wildlife

Some parts of the world are home to animals and plants that live nowhere else. These places are often remote islands, where life is cut off. In other cases, they are patches of unusual habitat, complete with the unique wildlife that depends on it.

California
A Mediterranean-type climate results in some unique forests featuring the world's largest living organism—the giant sequoia, a gigantic species of coniferous tree.

Mexican pine-oak forests
These forests on Mexican mountain ridges are patches of habitat not found anywhere else nearby. There are nearly 4,000 endemic plants and unique birds such as the Montezuma quail.

Hawaii and Polynesia
Only certain life-forms have reached these remote islands. Hawaii has no ants, but has 500 species of unique fruit flies, all evolved from a single species blown ashore 8 million years ago. Some of them are flightless and have taken up antlike lifestyles. Hawaii also has many unique plants, including the strange Hawaiian silversword, endemic to its mountaintops.

Galápagos Islands
These islands were made famous by Charles Darwin for their unique wildlife, including their giant tortoises.

Tropical Andes
Perhaps the richest region on Earth, these mountains are home to 664 species of amphibians, 450 of which are in danger of dying out. Of 1,700 bird species, 600—including this fiery-throated fruiteater—are found nowhere else.

Western Mediterranean
Europe's hot spot of unique wildlife. One species of midwife toad lives only on Majorca, and Barbary macaques live only on Gibraltar and in patches of habitat in Morocco and Algeria.

Canary Islands
Rich in endemic plants, the Canary Islands off Africa gave their name to the bird that lives only here and on nearby Atlantic islands—the canary.

Caribbean Islands
Each island has its own versions of many plants and animals. This Cuban knight anole lives only on Cuba.

Atlantic Forest
This thin strip of rainforest is cut off from the Amazon rainforest, so it has its own set of wildlife, including the endangered golden lion tamarin.

75 PERCENT OF THE UNIQUE PLANTS OF THE CANARY ISLANDS ARE ENDANGERED

NEARLY 7 PERCENT OF THE WORLD'S PLANTS ARE UNIQUE TO THE

ENDEMIC HOT SPOTS

Scientists have shown that these regions have the greatest number of plant species living only within a small area. They call these species "endemic" to that area. In these hot spots of unique plants, scientists tend to find lots of endemic animals, too.

Region rich in endemic species

BIOMES

- Tropical dry broad-leaved forest
- Tropical coniferous forest
- Temperate broad-leaved forest
- Temperate coniferous forest
- Tropical moist broad-leaved forest
- Boreal forest
- Savanna
- Flooded savanna
- Steppe
- Mountain grasslands and shrublands
- Mediterranean shrublands
- Desert and dry shrublands
- Arctic tundra
- Polar desert
- Mangroves

Mountains of southwest China
Each ridge of mountains has its own distinct wildlife. Endangered species, such as the Yunnan snub-nosed monkey, live only here.

Eastern Mediterranean
The Cedar of Lebanon lives only in a small area, including Lebanon, Israel, Palestine, and parts of Syria, Jordan, and Turkey.

Philippines
Of this country's 1,000 types of orchids, 70 percent grow nowhere else.

Wallacea
This region is named after 19th-century naturalist Alfred Russel Wallace, who noticed its unique wildlife such as the piglike babirusa.

Ethiopian Highlands
These highlands are home to 30 endemic bird species and the endangered Ethiopian wolf.

New Guinea
This large island is home to many unique birds of paradise and several endemic tree kangaroos, including this species, the ursine tree kangaroo.

Sri lanka and Western Ghats
This hot spot is home to 5,000 species of flowering plants, 139 mammal species, 508 birds, and 179 amphibian species.

East Melanesia
This string of islands has 3,000 endemic plant species and spectacular birdwing and swallowtail butterflies. This is a Ulysses swallowtail.

Madagascar
Ninety-eight percent of Madagascar's land mammals, 92 percent of its reptiles, 68 percent of its plants, and 41 percent of its breeding bird species exist nowhere else on Earth. All 16 mantella frogs are also endemic to the island.

Sundaland
Naturalists outline this region because its wildlife is distinct from next-door regions. One bizarre plant unique to Sundaland is *Rafflesia*, the stinking corpse lily.

East African Highlands
These islands of high ground in a sea of savanna support unusual plants such as this giant lobelia that grows on the slopes of Mount Kenya and Kilimanjaro.

Cape region
This is a small area of amazingly distinctive plantlife, including 6,000 endemic species such as this pincushion protea.

Western Australia
Like the South African Cape region, this is a "habitat island" of Mediterranean-type shrubland, full of plants found nowhere else, including the odd "kangaroo paw."

New Caledonia
Nothing like the strange, flightless kagu bird is found anywhere else in the world.

TROPICAL ANDES, WHICH COVER ONLY 0.8 PERCENT OF THE LAND AREA.

Kittlitz's murrelet
Alaska and Russian Far East

Maui parrotbill
Hawaii

Vaquita
Gulf of California

Hawaiian monk seal
Hawaii

Blue iguana
Grand Cayman Island, Caribbean

Iberian lynx
Spain

Lamotte's roundleaf bat
Mount Nimba (border area of Guinea, Liberia, and Côte d'Ivoire)

Variable harlequin frog
Costa Rica

Short-tailed chinchilla
Mountains on the Bolivia–Chile border

Maui parrotbill
In danger because of the loss of its forest habitat—only about 500 now survive.

Hawaiian monk seal
Once hunted for its skin and oil, today many become tangled in fishing nets or die because of pollution.

Glaucous macaw
Argentina, Uruguay, Paraguay, and Brazil

MORE THAN 7,000 ANIMAL SPECIES ARE CRITICALLY ENDANGERED

Western gorilla
Congo rainforest

Blue-eyed black lemur
Madagascar

In the red

Animals on the Red List—a list kept by the IUCN (International Union for the Conservation of Nature)—are in varying levels of endangerment. Those that are "critically endangered" may soon die out completely in the wild.

Vaquita This porpoise is the world's most endangered sea mammal; scientists estimate only about 10 are left.

Kittlitz's murrelet Thousands of these seabirds have been killed by sticky oil, spilled from giant tankers.

Blue iguana This lizard lives only on Grand Cayman Island. Numbers are increasing due to conservation.

Variable harlequin frog One of several harlequin frog species critically endangered due to a fungal disease.

Short-tailed chinchilla Hunted for its soft gray fur, this rock-dwelling rodent is now almost extinct in the wild.

Glaucous macaw Became rare because so many were caught and sold as pets. Only sighted twice in 100 years.

Iberian lynx If it dies out, it will be the first big cat species to go extinct in 10,000 years.

Western gorilla Many of these apes are killed for their meat, or have died from disease.

Lamotte's roundleaf bat This African mammal has become endangered mainly through the loss of its habitat.

Greater bamboo lemur Less than 100 have been spotted in 20 years of surveys.

Blue-eyed black lemur Like many other lemurs, this one could soon die out due to loss of its forest habitat.

Russian sturgeon This fish has been killed for its roe (eggs), known as caviar.

Indian vulture Many of these birds died after feeding on cattle that had been given drugs to help them work longer.

Bactrian camel Fewer than 1,000 survive in the wild.

Irrawaddy river shark As no one has seen this species for many years, it may be extinct in the wild.

Sumatran orangutan Just 15,000 of this species are left, since their forest is being cut down.

PEOPLE ARE WORKING HARD TO SAVE THE WORLD'S FEW REMAINING

Endangered animals

Our world has thousands of species, or kinds, of animals. Many are in danger of dying out, mainly because humans are destroying their habitats, or homes. Some animals have not been seen in their habitats for 50 years or more and can be declared "extinct in the wild."

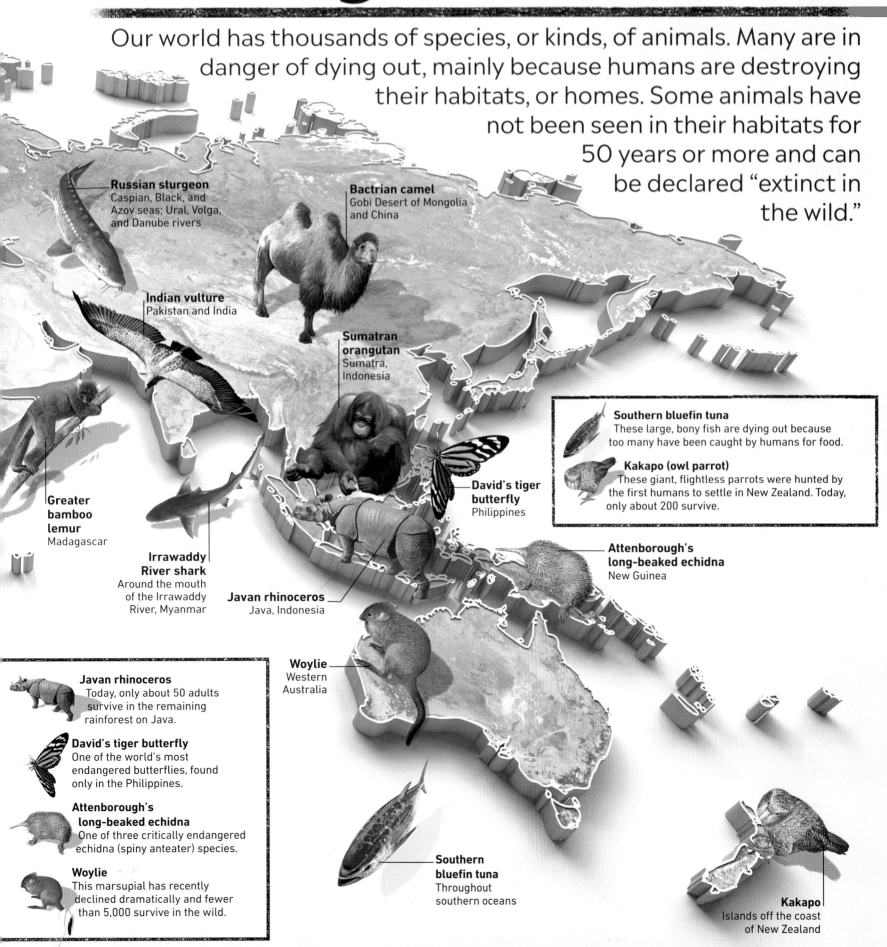

Russian sturgeon
Caspian, Black, and Azov seas; Ural, Volga, and Danube rivers

Bactrian camel
Gobi Desert of Mongolia and China

Indian vulture
Pakistan and India

Sumatran orangutan
Sumatra, Indonesia

Greater bamboo lemur
Madagascar

Irrawaddy River shark
Around the mouth of the Irrawaddy River, Myanmar

Javan rhinoceros
Java, Indonesia

David's tiger butterfly
Philippines

Attenborough's long-beaked echidna
New Guinea

Woylie
Western Australia

Southern bluefin tuna
Throughout southern oceans

Kakapo
Islands off the coast of New Zealand

Southern bluefin tuna
These large, bony fish are dying out because too many have been caught by humans for food.

Kakapo (owl parrot)
These giant, flightless parrots were hunted by the first humans to settle in New Zealand. Today, only about 200 survive.

Javan rhinoceros
Today, only about 50 adults survive in the remaining rainforest on Java.

David's tiger butterfly
One of the world's most endangered butterflies, found only in the Philippines.

Attenborough's long-beaked echidna
One of three critically endangered echidna (spiny anteater) species.

Woylie
This marsupial has recently declined dramatically and fewer than 5,000 survive in the wild.

KAKAPO—EVERY BIRD IS PROTECTED, AND EACH ONE HAS A NAME!

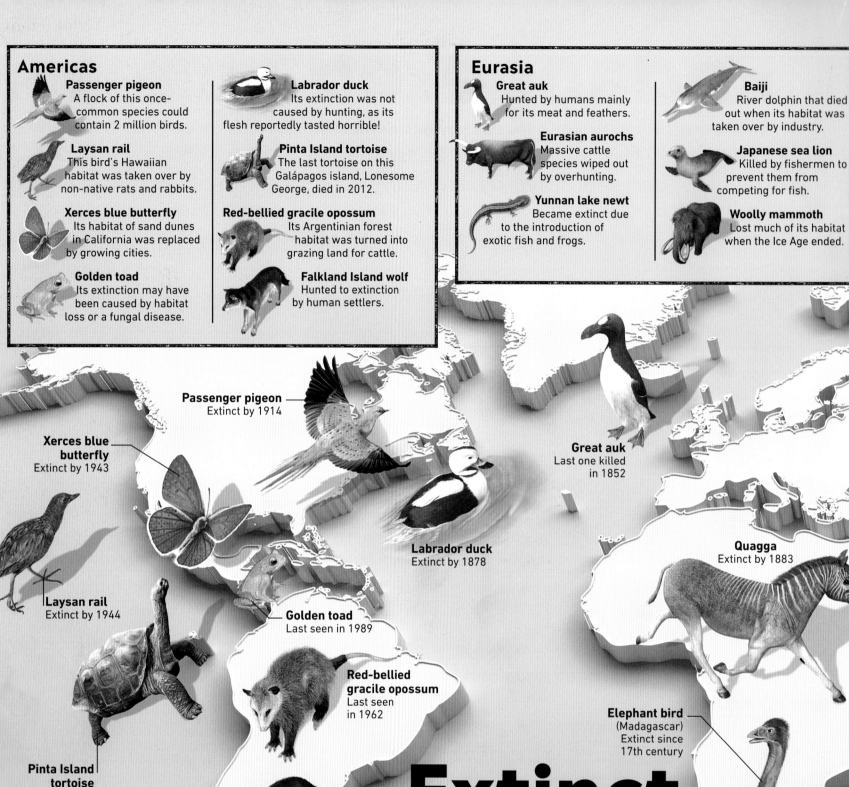

Americas

Passenger pigeon
A flock of this once-common species could contain 2 million birds.

Laysan rail
This bird's Hawaiian habitat was taken over by non-native rats and rabbits.

Xerces blue butterfly
Its habitat of sand dunes in California was replaced by growing cities.

Golden toad
Its extinction may have been caused by habitat loss or a fungal disease.

Labrador duck
Its extinction was not caused by hunting, as its flesh reportedly tasted horrible!

Pinta Island tortoise
The last tortoise on this Galápagos island, Lonesome George, died in 2012.

Red-bellied gracile opossum
Its Argentinian forest habitat was turned into grazing land for cattle.

Falkland Island wolf
Hunted to extinction by human settlers.

Eurasia

Great auk
Hunted by humans mainly for its meat and feathers.

Eurasian aurochs
Massive cattle species wiped out by overhunting.

Yunnan lake newt
Became extinct due to the introduction of exotic fish and frogs.

Baiji
River dolphin that died out when its habitat was taken over by industry.

Japanese sea lion
Killed by fishermen to prevent them from competing for fish.

Woolly mammoth
Lost much of its habitat when the Ice Age ended.

Passenger pigeon
Extinct by 1914

Xerces blue butterfly
Extinct by 1943

Laysan rail
Extinct by 1944

Golden toad
Last seen in 1989

Labrador duck
Extinct by 1878

Great auk
Last one killed in 1852

Quagga
Extinct by 1883

Red-bellied gracile opossum
Last seen in 1962

Pinta Island tortoise
Extinct in 2012

Elephant bird
(Madagascar) Extinct since 17th century

Falkland Island wolf
Presumed extinct in 1876

Extinct animals

The animal species on this map died out, or became extinct, quite recently and probably as a result of the actions of humans. But extinction has been happening naturally in the animal kingdom for millions of years.

ANIMALS ARE GOING EXTINCT TODAY AT LEAST X 100

Africa

Quagga
Its very distinctive markings made it an easy target for hunters.

Aldabra banded snail
A sudden decrease in rainfall, possibly caused by climate change, spelled extinction for this species.

Large sloth lemur
Gorilla-sized species that died out in Madagascar about 400 years ago.

Elephant bird
Huge flightless bird that was wiped out by hunting.

Dodo
This flightless bird became extinct within only 100 years of humans and their domestic animals arriving on the island of Mauritius.

Australasia

Lesser bilby
Probably wiped out by cats and foxes.

Eastern hare wallaby
Extinction was partly due to the introduction of cats, which hunted them.

Desert-rat kangaroo
Thought extinct, recovered, then declared extinct again in 1994.

King Island emu
Wiped out by sealers and their hunting dogs.

Tasmanian wolf
Hunted and trapped by human settlers in Tasmania—its last hiding place.

Moa
Victims of overhunting and loss of habitat.

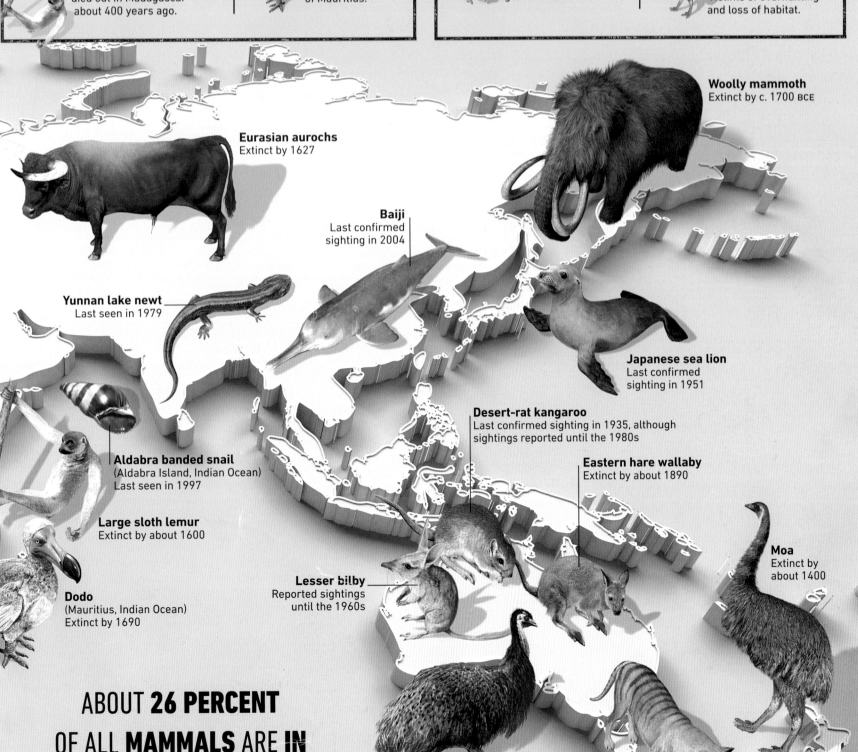

Woolly mammoth
Extinct by c. 1700 BCE

Eurasian aurochs
Extinct by 1627

Baiji
Last confirmed sighting in 2004

Yunnan lake newt
Last seen in 1979

Japanese sea lion
Last confirmed sighting in 1951

Desert-rat kangaroo
Last confirmed sighting in 1935, although sightings reported until the 1980s

Eastern hare wallaby
Extinct by about 1890

Aldabra banded snail
(Aldabra Island, Indian Ocean)
Last seen in 1997

Large sloth lemur
Extinct by about 1600

Dodo
(Mauritius, Indian Ocean)
Extinct by 1690

Lesser bilby
Reported sightings until the 1960s

Moa
Extinct by about 1400

ABOUT 26 PERCENT OF ALL MAMMALS ARE IN DANGER OF EXTINCTION

King Island emu
Extinct by around 1802

Tasmanian wolf
Presumed extinct in 1936

TIMES FASTER THAN THE NATURAL EXTINCTION RATE.

People and planet

Sprawling city
Los Angeles, California, stretches as far as the horizon in this photo taken from Mount Hollywood. The skyscrapers of downtown LA can be seen on the left.

Introduction

Humans, together with animals and other living things, form what is called the biosphere—the living part of the world. Since modern humans first appeared in Africa about 200,000 years ago, we have colonized virtually the entire world—even hot deserts and the ice-cold Arctic. As we have done so, our impact on the biosphere has been far-reaching.

Human impact

The human "footprint" on planet Earth is deep and broad. We have transformed the landscape—clearing forests to produce food, digging minerals and ores from the ground, and channeling and storing water to meet our needs. Our living space is concentrated into larger and larger cities, but these cities are hungry for food and energy taken from the surrounding land.

Renewable energy
New ways of harnessing the energy of sunlight and wind are reducing our use of fossil fuels. Unlike fossil fuels, these energy sources will never run out.

Natural resources

Buried within Earth's crust there are limited supplies of minerals, metal ores, and fossil fuels (coal, oil, and gas). Once these reserves are exhausted, they cannot be replaced. Burning these fuels also damages Earth's atmosphere and is contributing to global warming.

Population

For most of humanity's existence, the human population grew relatively slowly. In 10,000 BCE, there were only 1–5 million people on Earth. By 1000 BCE, after farming was invented, the population had increased to about 50 million. Since reaching the 1 billion mark in 1804, during the early Industrial Revolution, the population has expanded much more quickly than ever before.

Growing bigger, fast
The period since the late 1950s has seen the human population more than double.

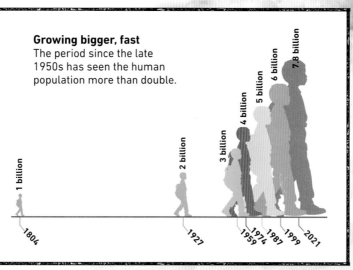

1 billion · 1804
2 billion · 1927
3 billion · 1959
4 billion · 1974
5 billion · 1987
6 billion · 1999
7.8 billion · 2021

Agriculture
In 1700 CE, about 7 percent of Earth's land area was used for growing crops and raising farm animals. Today, that figure has risen to about 50 percent.

Pollution
Vehicle exhaust gases, smoke and waste chemicals from factories, and oil spills all poison the environment, threatening plant and animal life.

Conservation
To protect the plant and animal life of unique habitats, many countries set up conservation areas, where no farming, industry, or new settlement can occur.

Using water
We build dams and reservoirs to store water. We need it for drinking, for use in industrial processes, and to irrigate crops and generate electricity.

Successful species

Part of the success of humans is due to our ability to use the materials around us to give us protection and shelter. This ability opens up nearly every part of the globe for human living space, no matter how harsh the environment. Even thousands of years ago, the Inuit of the North American Arctic made coats from the fur of caribou (reindeer) and waterproof boots from seal skin. They found a way to live on the meager resources of the high Arctic.

Inuit boat
The *umiak* is a type of traditional open boat used by Inuit people. The frame is made of driftwood or whalebone, with a walrus- or seal-skin covering. These boats are still used, since the law allows whale hunting only with traditional Inuit tools.

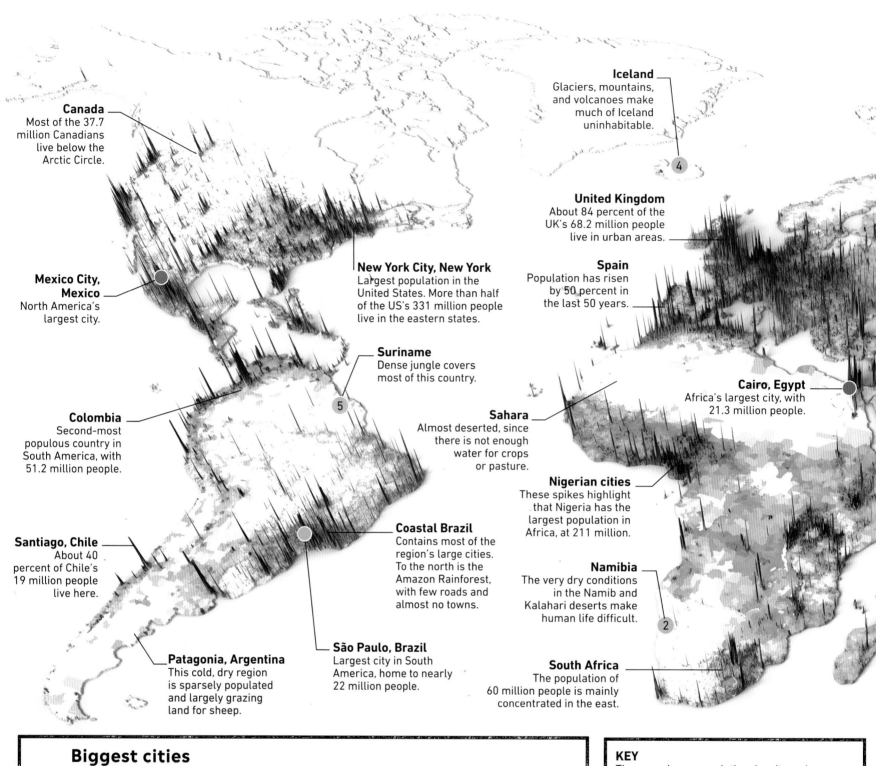

Canada
Most of the 37.7 million Canadians live below the Arctic Circle.

Iceland
Glaciers, mountains, and volcanoes make much of Iceland uninhabitable.
4

United Kingdom
About 84 percent of the UK's 68.2 million people live in urban areas.

Mexico City, Mexico
North America's largest city.

New York City, New York
Largest population in the United States. More than half of the US's 331 million people live in the eastern states.

Spain
Population has risen by 50 percent in the last 50 years.

Suriname
Dense jungle covers most of this country.
5

Cairo, Egypt
Africa's largest city, with 21.3 million people.

Colombia
Second-most populous country in South America, with 51.2 million people.

Sahara
Almost deserted, since there is not enough water for crops or pasture.

Nigerian cities
These spikes highlight that Nigeria has the largest population in Africa, at 211 million.

Coastal Brazil
Contains most of the region's large cities. To the north is the Amazon Rainforest, with few roads and almost no towns.

Namibia
The very dry conditions in the Namib and Kalahari deserts make human life difficult.
2

Santiago, Chile
About 40 percent of Chile's 19 million people live here.

Patagonia, Argentina
This cold, dry region is sparsely populated and largely grazing land for sheep.

São Paulo, Brazil
Largest city in South America, home to nearly 22 million people.

South Africa
The population of 60 million people is mainly concentrated in the east.

Biggest cities

More than half the world's people now live in towns and cities, rather than in the countryside. Many cities have grown quickly and have been dubbed "megacities," with more than 10 million people in a metropolitan area. Below are the 10 largest.

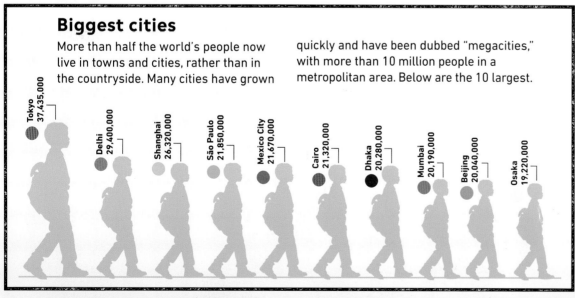

Tokyo 37,435,000
Delhi 29,400,000
Shanghai 26,320,000
São Paulo 21,850,000
Mexico City 21,670,000
Cairo 21,320,000
Dhaka 20,280,000
Mumbai 20,190,000
Beijing 20,040,000
Osaka 19,220,000

KEY
The map shows population density, or how closely people are packed together. Denser places, such as cities, appear as red mountains.

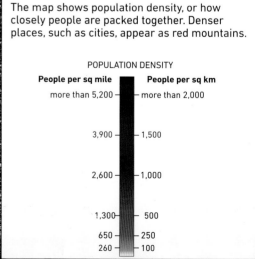

POPULATION DENSITY

People per sq mile	People per sq km
more than 5,200	more than 2,000
3,900	1,500
2,600	1,000
1,300	500
650	250
260	100

IN 1800, THE WORLD'S POPULATION WAS LESS THAN 1 BILLION PEOPLE

Where people live

The world's 7.8 billion people are not spread evenly across the globe: most live where there are natural resources and fertile land for farming. Some places are too hostile for humans to thrive.

Siberia, Russia
Few people live here, since the climate is too cold to grow crops. Some spikes show the location of cities based around extracting oil and gas from under the frozen tundra.

Moscow, Russia
Home to 12.5 million people.

Kolkata, India
Center of eastern India.

Mongolia
Little of the land is good for growing crops and many people are scattered in small communities of nomadic herdspeople.

Shanghai, China
China's largest city.

Beijing
The capital of China.

Tokyo, Japan
The largest city in the world since the 1960s.

Osaka, Japan
The second-largest city in Japan.

Delhi, India
India's capital sits in the densely populated Ganges River basin, home to 650 million people packed in at nearly 1,000 per sq mile (400 per sq km).

Eastern China
Most of China's 1.4 billion people live here.

Manila, Philippines
Not including its outlying districts, this is the world's most densely populated city.

Dhaka, Bangladesh
The world's most densely populated, continuously built-up area.

Mumbai, India
Fast-growing entertainment hub of India.

Jakarta, Indonesia
Of all Indonesia's islands, Java is by far the most crowded and contains the booming capital, Jakarta.

IN **MANILA**, PHILIPPINES, ON AVERAGE **296 PEOPLE LIVE** IN AN AREA THE SIZE OF A **SOCCER PITCH**

Australia
Australia's center is too dry to support farming and very few people live here.

Auckland, New Zealand
About one in three New Zealanders live here.

Melbourne, Australia
Most of Australia's population lives on the southeastern coast, in cities including Melbourne.

Most sparsely populated countries

		total population	people per sq mile	people per sq km
1	Mongolia	3,278,000	5.5	2.1
2	Namibia	2,541,000	8.0	3.1
3	Australia	25,500,000	8.6	3.3
4	Iceland	341,000	8.8	3.4
5	Suriname	587,000	9.7	3.8

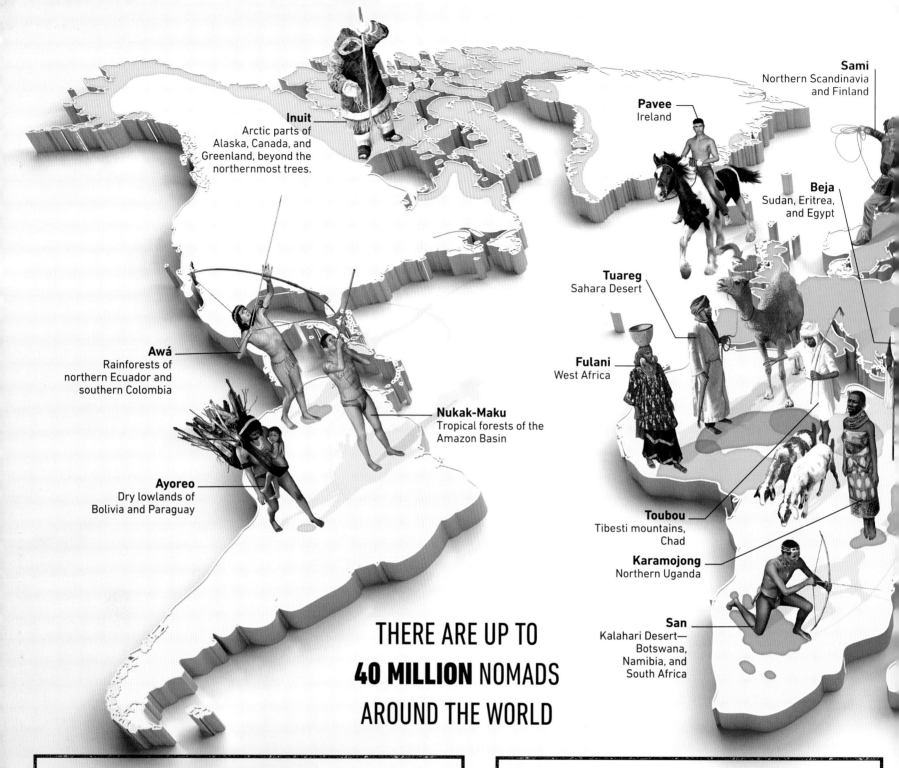

Inuit
Arctic parts of Alaska, Canada, and Greenland, beyond the northernmost trees.

Sami
Northern Scandinavia and Finland

Pavee
Ireland

Beja
Sudan, Eritrea, and Egypt

Awá
Rainforests of northern Ecuador and southern Colombia

Nukak-Maku
Tropical forests of the Amazon Basin

Tuareg
Sahara Desert

Fulani
West Africa

Ayoreo
Dry lowlands of Bolivia and Paraguay

Toubou
Tibesti mountains, Chad

Karamojong
Northern Uganda

San
Kalahari Desert— Botswana, Namibia, and South Africa

THERE ARE UP TO
40 MILLION NOMADS
AROUND THE WORLD

Americas

Inuit
For 4,000 years, the Inuit have roamed the region they call Nunavut, "our land."

Awá
The Awá speak their own ancient language called Awa Pit.

Nukak-Maku
The Nukak people are expert hunters who were entirely isolated until 1988.

Ayoreo
The Ayoreo mix a hunter-gatherer lifestyle with agriculture.

Europe

Pavee, or Irish Travelers
The Pavee have strict moral beliefs laid out in "The Travelers' Code."

Sami
The Sami reindeer herders and fur trappers have existed for over 5,000 years.

Roma
There are 2–5 million Roma worldwide, mostly in Europe.

Nenets
Every year, Nenets move huge herds of reindeer up to 620 miles (1,000 km).

Africa

Beja
Only some Beja clans are nomadic.

Tuareg
In Tuareg culture, men rather than women wear the veil.

Toubou
The Toubou are divided into two peoples: the Teda and the Daza.

Fulani
The Fulani traditionally herd goats, sheep, and cattle across large areas of west Africa.

Gabra
These herders make their dome-shaped houses out of acacia roots and cloth.

Afar
The Afar live by rivers in the dry season and head for higher ground in the wet season.

Karamojong
This name means "the old men can walk no further."

San
The San are famous for being excellent trackers and hunters.

MOST NOMADS LIVE IN DESERT, STEPPE, OR TUNDRA—DRY PLACES THAT

Roma
Central and eastern Europe

Kazakhs
Kazhakstan and other parts of northern central Asia

Nenets
Arctic Russia

Asia

Bakhtiari
Bakhtiari means "bearer of good luck." Some still move pastures with the seasons.

Kazakhs
There are still many nomadic Kazakhs left in Xinjiang, China.

Bedouin
Bedouin are desert-dwelling wanderers known for their hospitality.

Qashqai
Qashqai are traditionally farmers known for their beautiful wool products.

Bakhtiari
Southwestern Iran

Yakut
The Yakutia Republic, Russia

Chukchi
The Bering Strait region of Siberia

Qashqai
Southwestern Iran

Evenks
Southern Siberia, Mongolia, and northeasternmost China

Moken
Southern Burma and the west coast of Thailand

Bedouin
The Middle East, predominantly Saudi Arabia

Afar
The Horn of Africa

Gabra
Chalbi Desert of Kenya and highlands of southern Ethiopia

Penan
Sarawak, Malaysia

Yakut
The Yakut are seminomadic reindeer herders.

Evenks
The Evenks kept small herds of domesticated reindeer, which helped the people move around easily.

Chukchi
The word "chukchi" means "rich in reindeer."

Moken
Moken children have extremely good underwater vision due to diving for food.

Penan
In Penan society everything is shared.

Aboriginal peoples
Australia

Australasia

Aboriginal peoples
Groups of Aboriginals have lived all over Australia for about 60,000 years.

Nomads

Nomads move home every year to find fresh pasture or hunting grounds. Some are herders, some hunter-gatherers, and others are wandering traders. Their nomadic lifestyle is quickly dying out, as many of them are settling in villages and towns.

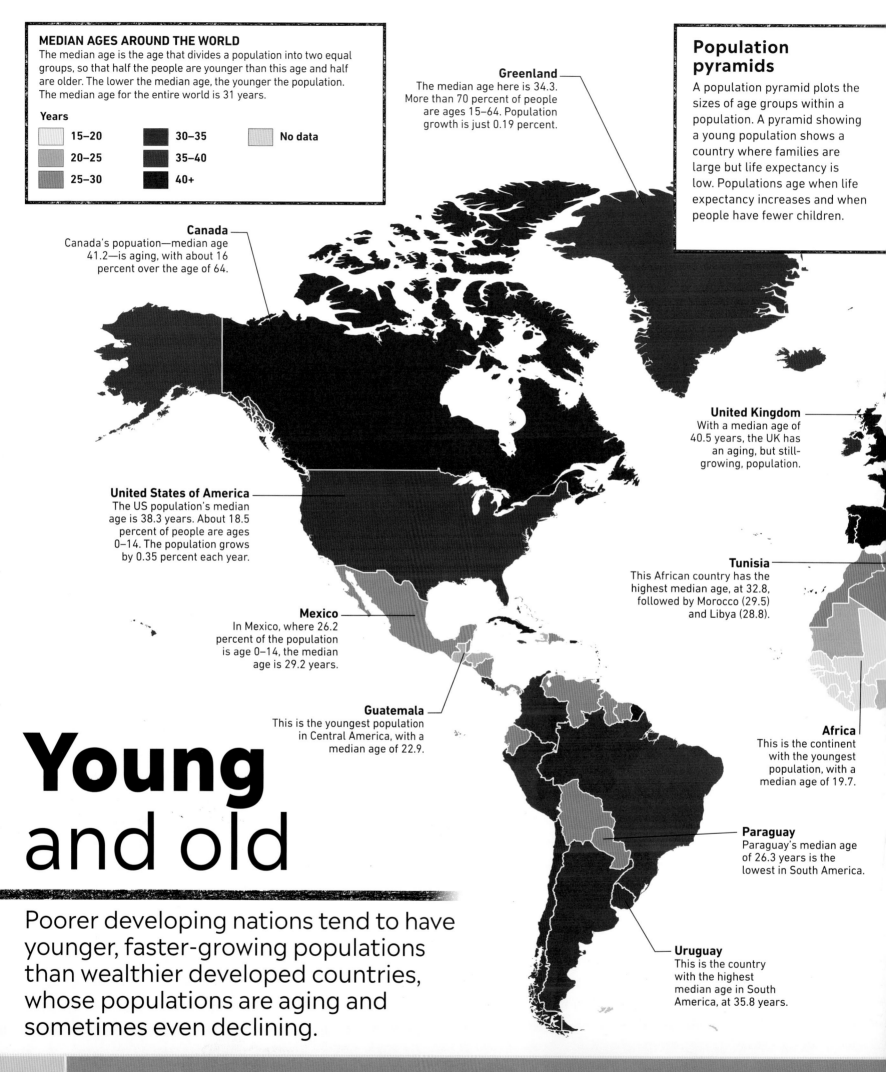

MEDIAN AGES AROUND THE WORLD

The median age is the age that divides a population into two equal groups, so that half the people are younger than this age and half are older. The lower the median age, the younger the population. The median age for the entire world is 31 years.

Years

15–20	30–35	No data
20–25	35–40	
25–30	40+	

Greenland
The median age here is 34.3. More than 70 percent of people are ages 15–64. Population growth is just 0.19 percent.

Population pyramids

A population pyramid plots the sizes of age groups within a population. A pyramid showing a young population shows a country where families are large but life expectancy is low. Populations age when life expectancy increases and when people have fewer children.

Canada
Canada's popuation—median age 41.2—is aging, with about 16 percent over the age of 64.

United Kingdom
With a median age of 40.5 years, the UK has an aging, but still-growing, population.

United States of America
The US population's median age is 38.3 years. About 18.5 percent of people are ages 0–14. The population grows by 0.35 percent each year.

Tunisia
This African country has the highest median age, at 32.8, followed by Morocco (29.5) and Libya (28.8).

Mexico
In Mexico, where 26.2 percent of the population is age 0–14, the median age is 29.2 years.

Guatemala
This is the youngest population in Central America, with a median age of 22.9.

Africa
This is the continent with the youngest population, with a median age of 19.7.

Paraguay
Paraguay's median age of 26.3 years is the lowest in South America.

Uruguay
This is the country with the highest median age in South America, at 35.8 years.

Young
and old

Poorer developing nations tend to have younger, faster-growing populations than wealthier developed countries, whose populations are aging and sometimes even declining.

BY 2050, ABOUT 16 PERCENT OF THE WORLD'S POPULATION WILL

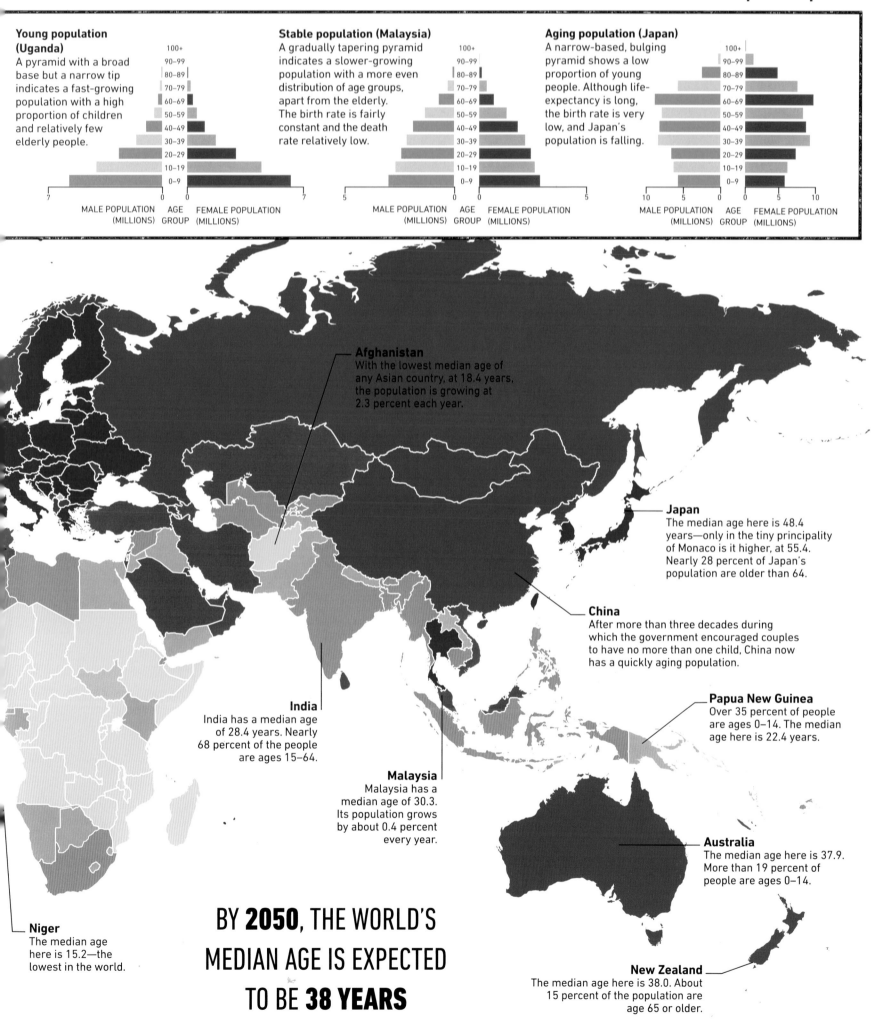

Young population (Uganda)

A pyramid with a broad base but a narrow tip indicates a fast-growing population with a high proportion of children and relatively few elderly people.

100+
90–99
80–89
70–79
60–69
50–59
40–49
30–39
20–29
10–19
0–9

7   0 0   7
MALE POPULATION   AGE   FEMALE POPULATION
(MILLIONS)   GROUP   (MILLIONS)

Stable population (Malaysia)

A gradually tapering pyramid indicates a slower-growing population with a more even distribution of age groups, apart from the elderly. The birth rate is fairly constant and the death rate relatively low.

100+
90–99
80–89
70–79
60–69
50–59
40–49
30–39
20–29
10–19
0–9

5   0 0   5
MALE POPULATION   AGE   FEMALE POPULATION
(MILLIONS)   GROUP   (MILLIONS)

Aging population (Japan)

A narrow-based, bulging pyramid shows a low proportion of young people. Although life-expectancy is long, the birth rate is very low, and Japan's population is falling.

100+
90–99
80–89
70–79
60–69
50–59
40–49
30–39
20–29
10–19
0–9

10   5   0 0   5   10
MALE POPULATION   AGE   FEMALE POPULATION
(MILLIONS)   GROUP   (MILLIONS)

Afghanistan
With the lowest median age of any Asian country, at 18.4 years, the population is growing at 2.3 percent each year.

Japan
The median age here is 48.4 years—only in the tiny principality of Monaco is it higher, at 55.4. Nearly 28 percent of Japan's population are older than 64.

China
After more than three decades during which the government encouraged couples to have no more than one child, China now has a quickly aging population.

Papua New Guinea
Over 35 percent of people are ages 0–14. The median age here is 22.4 years.

India
India has a median age of 28.4 years. Nearly 68 percent of the people are ages 15–64.

Malaysia
Malaysia has a median age of 30.3. Its population grows by about 0.4 percent every year.

Australia
The median age here is 37.9. More than 19 percent of people are ages 0–14.

Niger
The median age here is 15.2—the lowest in the world.

New Zealand
The median age here is 38.0. About 15 percent of the population are age 65 or older.

BY **2050**, THE WORLD'S MEDIAN AGE IS EXPECTED TO BE **38 YEARS**

BE AGE 65 OR OVER, COMPARED WITH 7.6 PERCENT IN 2010.

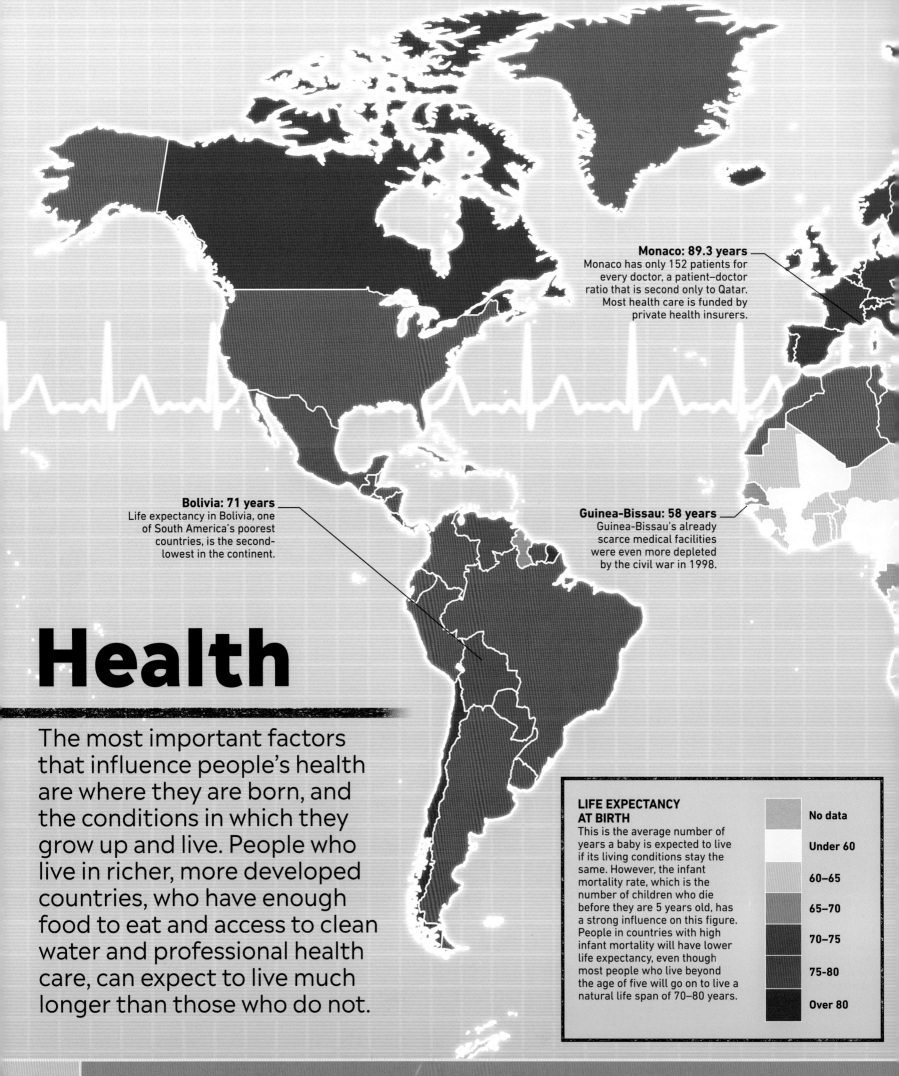

Monaco: 89.3 years
Monaco has only 152 patients for every doctor, a patient–doctor ratio that is second only to Qatar. Most health care is funded by private health insurers.

Bolivia: 71 years
Life expectancy in Bolivia, one of South America's poorest countries, is the second-lowest in the continent.

Guinea-Bissau: 58 years
Guinea-Bissau's already scarce medical facilities were even more depleted by the civil war in 1998.

Health

The most important factors that influence people's health are where they are born, and the conditions in which they grow up and live. People who live in richer, more developed countries, who have enough food to eat and access to clean water and professional health care, can expect to live much longer than those who do not.

LIFE EXPECTANCY AT BIRTH
This is the average number of years a baby is expected to live if its living conditions stay the same. However, the infant mortality rate, which is the number of children who die before they are 5 years old, has a strong influence on this figure. People in countries with high infant mortality will have lower life expectancy, even though most people who live beyond the age of five will go on to live a natural life span of 70–80 years.

	No data
	Under 60
	60–65
	65–70
	70–75
	75–80
	Over 80

A CHILD BORN IN ESWATINI IS NEARLY 30 TIMES MORE LIKELY

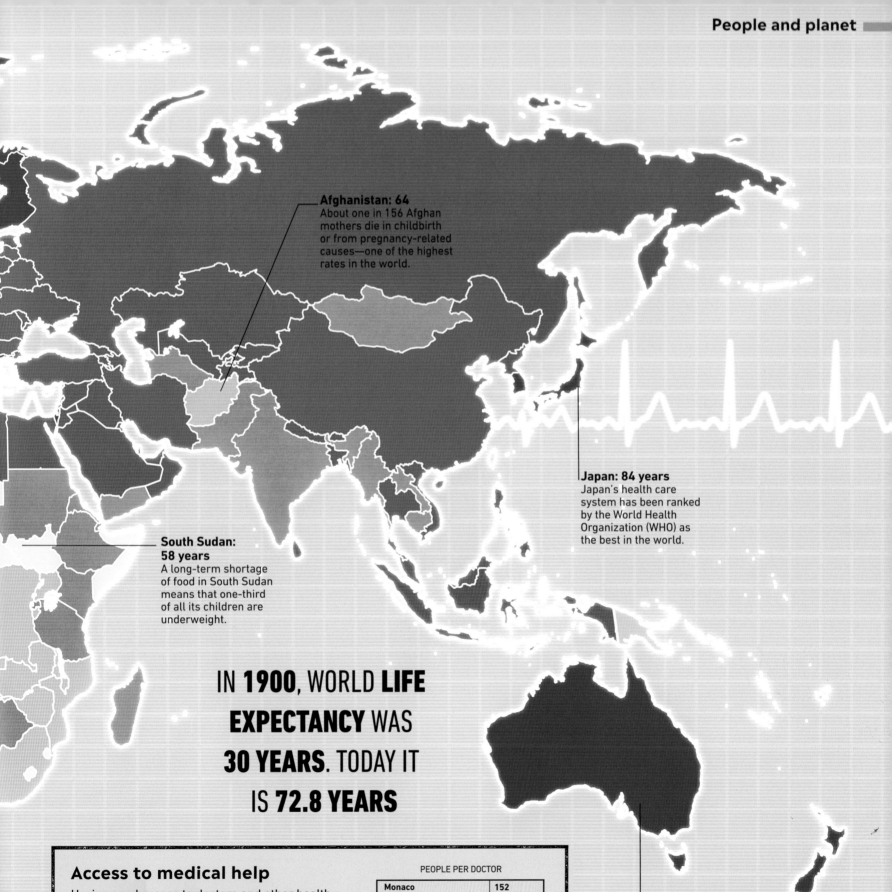

Afghanistan: 64
About one in 156 Afghan mothers die in childbirth or from pregnancy-related causes—one of the highest rates in the world.

Japan: 84 years
Japan's health care system has been ranked by the World Health Organization (WHO) as the best in the world.

South Sudan: 58 years
A long-term shortage of food in South Sudan means that one-third of all its children are underweight.

IN **1900**, WORLD **LIFE EXPECTANCY** WAS **30 YEARS**. TODAY IT IS **72.8 YEARS**

Australia: 83 years
Life expectancy among the Aboriginal population of Australia is only 73.6 years, much lower than the national average.

Access to medical help

Having good access to doctors and other health-care workers is essential in helping people to stay healthy, recover from illness, and live longer. The number of doctors per person in the population has an important effect on life expectancy, but other factors influence people's life span. Monaco, for instance, has roughly the same number of doctors per head as Cuba, but life expectancy in Monaco is over ten years longer than that in Cuba.

PEOPLE PER DOCTOR	
Monaco	152
Cuba	149
St Lucia	204
Belarus	254
Georgia	234
Liberia	15,000
Mozambique	33,300
Niger	50,000
Bhutan	3,846
Malawi	50,000

TO DIE BEFORE THE AGE OF FIVE THAN A CHILD BORN IN SWEDEN.

Infecting germs

Many infectious diseases are caused by microscopic living organisms. They live and multiply inside our bodies and can pass from human to human by touch, through blood or saliva, and through the air.

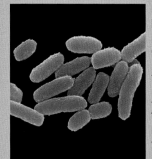

Bubonic plague bacteria

Bacteria are single-celled organisms that multiply by dividing into two again and again. Millions could fit on the head of a pin. Today, many bacterial infections can be treated with antibiotics.

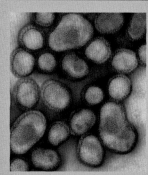

Flu virus

Viruses are very simple organisms far smaller even than bacteria. They spread by invading and taking over cells in the body. Viruses are unharmed by antibiotics, but the body can be fortified against them with a vaccine.

The Black Death ravaged Britain in 1348–50.

Troops returning home from Asia at the end of World War I brought the Spanish Flu back with them.

In August 1918, a second wave of Spanish Flu crossed the Atlantic and hit the port city of Freetown, Sierra Leone.

③

Spanish Flu

This infection was called "Spanish Flu" because people first thought it began in Spain. However, it actually was first reported at a training camp for American soldiers in the United States. The disease spread quickly when infected soldiers traveled to Europe to fight in World War I. It is estimated to have killed 20–50 million people.

④

Freetown

According to some studies, HIV began its spread through the human population in Cameroon.

Pandemics

Infectious diseases—illnesses that pass between people— can spread rapidly. Many people become ill, causing a local disaster called an epidemic. When this effect becomes global, we call it a pandemic.

KEY

This map shows the spread of three of history's most lethal pandemics—in ancient times, the Middle Ages, and modern times.

 Plague of Justinian
Bubonic plague, 541–42 CE

 Black Death
Bubonic plague, 1346–55 CE

 Spanish Flu
Influenza, 1918–20

SPANISH FLU MAY HAVE KILLED UP TO 50 MILLION PEOPLE.

Superbugs and new viruses

Bacteria and viruses change fast. "Superbug" bacteria become immune to antibiotics, while scientists try to develop vaccines against new viruses. Today, air travel can spread infection worldwide in days, so the fear of a fast-spreading pandemic is greater than ever. Here are five recent cases of new viruses.

1 Hong Kong Flu, 1968–69
In 2 years, Hong Kong Flu caused about 1 million deaths. The virus killed about 34,000 people in the United States alone.

2 Avian (Bird) Flu, Hong Kong, 1997–present
This virus first appeared in humans in Hong Kong, through contact with infected poultry. It has killed hundreds of people since then.

3 H1N1 ("Swine Flu"), Mexico City, 2009–10
This new flu developed from viruses of birds, pigs, and humans. Up to 575,400 people died in the first year of this pandemic.

4 HIV, west–central Africa, 1981–present
This virus causes AIDS—an often-fatal disease of the body's defenses. It now infects more than 30 million people worldwide.

Black Death
In the 14th century, an outbreak of bubonic plague spread from Asia across Europe, causing devastation along the way. It caused some 50 million deaths—about half in Europe, where 25 percent of the population was killed.

Constantinople (Istanbul)

Some experts think the Plague of Justinian began not in Ethiopia, but in Central Asia.

The Black Death passed along sea trade routes, since the bacteria that caused the disease lived in fleas, which lived on ships' rats.

Plague of Justinian
At its height, during the rule of the Emperor Justinian (ruler of the Byzantine, or Eastern Roman, Empire), this disease killed at least 25 million people. It may have started in Ethiopia, then spread along trade routes through northern Egypt and Constantinople (modern-day Istanbul) into Europe.

COVID-19

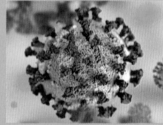

5 Dec 2019–present
First identified in Wuhan, China, in late 2019, this fast-spreading virus can cause severe respiratory problems; up to 2.6 million deaths were reported in the first year of the pandemic. Vaccines have now been developed to help protect against the disease.

UP TO **2,000 PEOPLE** STILL SUFFER FROM **PLAGUE** EACH YEAR

Spanish Flu was brought to New Zealand in 1918 by soldiers returning home from fighting in World War I in Europe.

THAT'S MORE THAN THE TOTAL NUMBER OF DEATHS DURING WORLD WAR I.

The poverty line

A poverty line is the minimum level of income thought to be enough for a person to live on. It is the least amount needed to provide basic necessities: food, clothing, health care, and shelter. The cost of living is different around the world, so the poverty line varies from country to country.

PEOPLE ON LESS THAN $1.90 A DAY

The international extreme poverty line of $1.90 income a day is a global measure of absolute poverty. This amount was set by the World Bank in 2015, and will be updated when necessary to reflect the cost of living. The map shows the percentage of each country's people earning less than $1.90 a day.

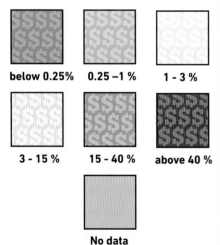

below 0.25%	0.25 –1 %	1 - 3 %
3 - 15 %	15 - 40 %	above 40 %

No data

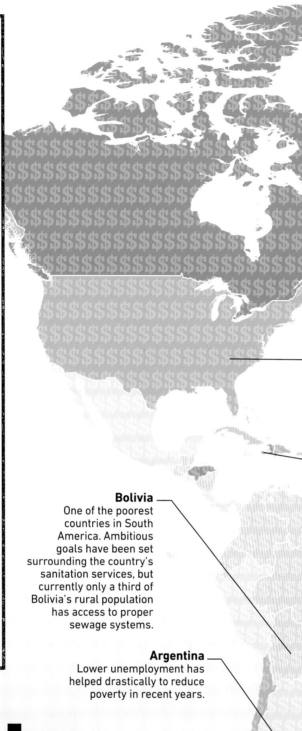

Morocco
Income inequality here is the highest in North Africa.

US
The wealth gap is huge in America; the top 1 percent of US households hold 15 times more wealth than the entirety of the lower 50 percent.

Haiti
The most cases of extreme poverty in the western hemisphere. Haiti's economy was severely affected by a 2010 earthquake, and is still yet to recover.

Bolivia
One of the poorest countries in South America. Ambitious goals have been set surrounding the country's sanitation services, but currently only a third of Bolivia's rural population has access to proper sewage systems.

Argentina
Lower unemployment has helped drastically to reduce poverty in recent years.

Liberia
One of the poorest countries in the world. An estimated 64 percent of the population lives below the $1.90-a-day line.

Ghana
While the overall poverty rate has gone down sharply over the last 30 years, poverty in the north of the country has changed little.

Poverty

The COVID-19 pandemic means that global poverty is expected to rise for the first time since 2000. Sub-Saharan Africa has by far the most cases of extreme poverty—half of the countries in this region have a poverty rate higher than 35%.

Inequality

In many countries, the gap between rich and poor is widening. Tax, special benefits for the lowest earners, and free education, among other things, can help reduce this. These charts show how much of a country's overall wealth the richest people own. The countries shown here are those with a very large gap between rich and poor, and those where the gap is less noticeable.

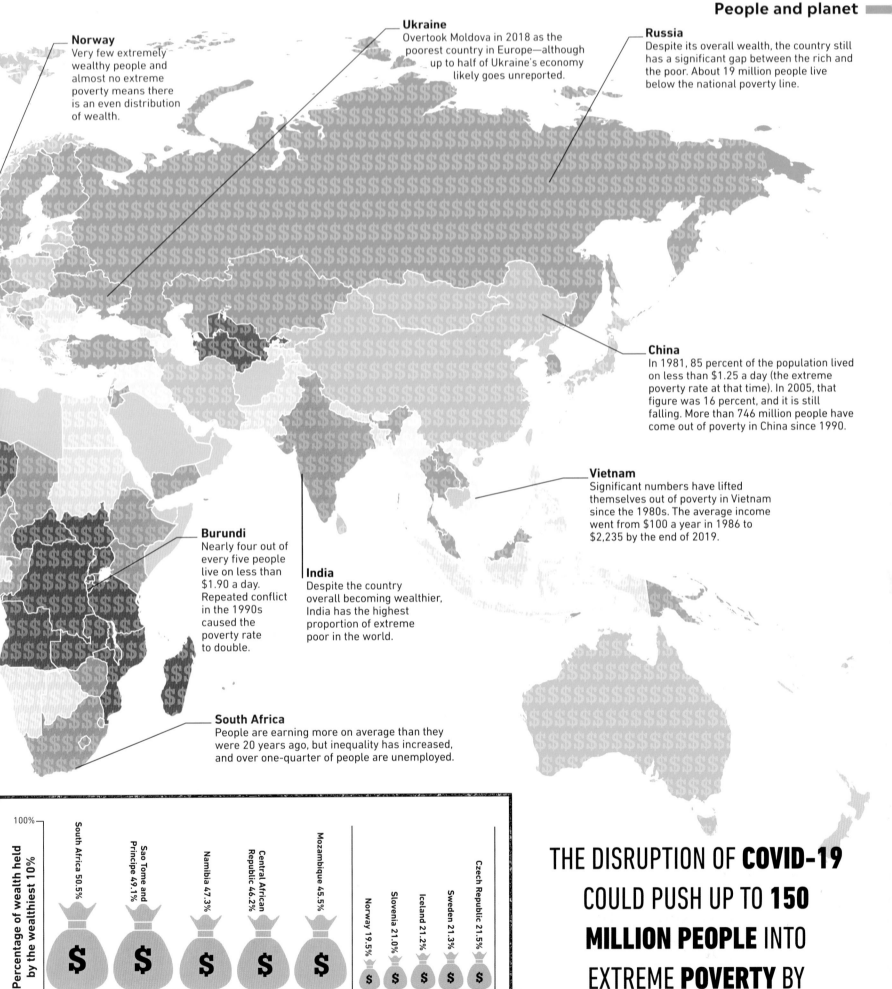

Norway
Very few extremely wealthy people and almost no extreme poverty means there is an even distribution of wealth.

Ukraine
Overtook Moldova in 2018 as the poorest country in Europe—although up to half of Ukraine's economy likely goes unreported.

Russia
Despite its overall wealth, the country still has a significant gap between the rich and the poor. About 19 million people live below the national poverty line.

China
In 1981, 85 percent of the population lived on less than $1.25 a day (the extreme poverty rate at that time). In 2005, that figure was 16 percent, and it is still falling. More than 746 million people have come out of poverty in China since 1990.

Vietnam
Significant numbers have lifted themselves out of poverty in Vietnam since the 1980s. The average income went from $100 a year in 1986 to $2,235 by the end of 2019.

Burundi
Nearly four out of every five people live on less than $1.90 a day. Repeated conflict in the 1990s caused the poverty rate to double.

India
Despite the country overall becoming wealthier, India has the highest proportion of extreme poor in the world.

South Africa
People are earning more on average than they were 20 years ago, but inequality has increased, and over one-quarter of people are unemployed.

Percentage of wealth held by the wealthiest 10%

100%

South Africa 50.5%
Sao Tome and Principe 49.1%
Namibia 47.3%
Central African Republic 46.2%
Mozambique 45.5%

Norway 19.5%
Slovenia 21.0%
Iceland 21.2%
Sweden 21.3%
Czech Republic 21.5%

Most unequal

Most equal

THE DISRUPTION OF **COVID-19** COULD PUSH UP TO **150 MILLION PEOPLE** INTO EXTREME **POVERTY** BY THE END OF 2021

The world's gold

Beautiful, rare, and highly prized, gold has been mined since ancient Egyptian times. Sometimes a discovery of gold led to a "gold rush," with thousands of people flocking to the site in the hope of making their fortune.

Klondike gold rush, Canada, 1897–99
100,000 prospectors headed for Klondike. About 4,000 found gold.

California gold rush, 1848–55
300,000 people flocked to California, aiming to strike gold.

Canada
Five percent of the world's gold comes from Canada.

United States
This is the fourth-largest gold producer, mining 220 tons annually (6 percent of the global total).

Ghana
Ghana is Africa's largest gold producer, snatching this top ranking from South Africa in 2019.

Peru
Peru is the largest gold producer in South America, and the world's sixth largest (4 percent of all gold).

Top 10 gold mines
Figures show gold mined in 2019.

1. **Muruntau, Uzbekistan**
   68.6 tons
2. **Olimpiada, Russia**
   47.6 tons
3. **Carlin, Nevada**
   45 tons
4. **Pueblo Viejo, Dominican Republic**
   33.7 tons
5. **Cortez, Nevada**
   32.9 tons
6. **Lihir, Papua New Guinea**
   30.2 tons
7. **Cadia East, Australia**
   29.8 tons
8. **Grasberg, Indonesia**
   29.5 tons
9. **Kibali, Democratic Republic of Congo**
   27.9 tons
10. **Loulo-Gounkoto, Mali**
   24.5 tons

ALL THE **GOLD** THAT HAS EVER BEEN **MINED** WOULD MAKE A CUBE **92 FT** (28 M) ALONG EACH SIDE

FOUND IN AUSTRALIA IN 1869, "WELCOME STRANGER" WAS THE BIGGEST

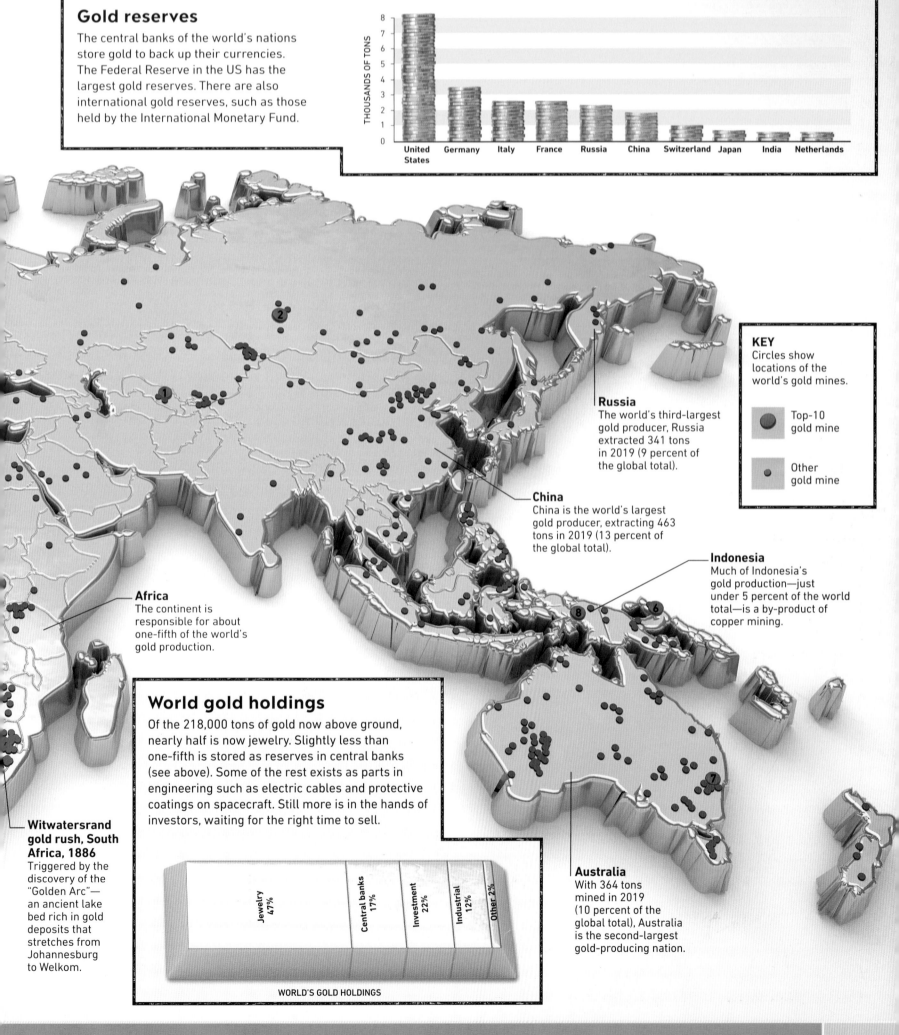

Gold reserves

The central banks of the world's nations store gold to back up their currencies. The Federal Reserve in the US has the largest gold reserves. There are also international gold reserves, such as those held by the International Monetary Fund.

THOUSANDS OF TONS

United States | Germany | Italy | France | Russia | China | Switzerland | Japan | India | Netherlands

KEY
Circles show locations of the world's gold mines.

Top-10 gold mine

Other gold mine

Russia
The world's third-largest gold producer, Russia extracted 341 tons in 2019 (9 percent of the global total).

China
China is the world's largest gold producer, extracting 463 tons in 2019 (13 percent of the global total).

Indonesia
Much of Indonesia's gold production—just under 5 percent of the world total—is a by-product of copper mining.

Africa
The continent is responsible for about one-fifth of the world's gold production.

World gold holdings

Of the 218,000 tons of gold now above ground, nearly half is now jewelry. Slightly less than one-fifth is stored as reserves in central banks (see above). Some of the rest exists as parts in engineering such as electric cables and protective coatings on spacecraft. Still more is in the hands of investors, waiting for the right time to sell.

Witwatersrand gold rush, South Africa, 1886
Triggered by the discovery of the "Golden Arc"—an ancient lake bed rich in gold deposits that stretches from Johannesburg to Welkom.

Jewelry 47% | Central banks 17% | Investment 22% | Industrial 12% | Other 2%

WORLD'S GOLD HOLDINGS

Australia
With 364 tons mined in 2019 (10 percent of the global total), Australia is the second-largest gold-producing nation.

GOLD NUGGET EVER FOUND. IT CONTAINED 156 LB (71 KG) OF GOLD.

89

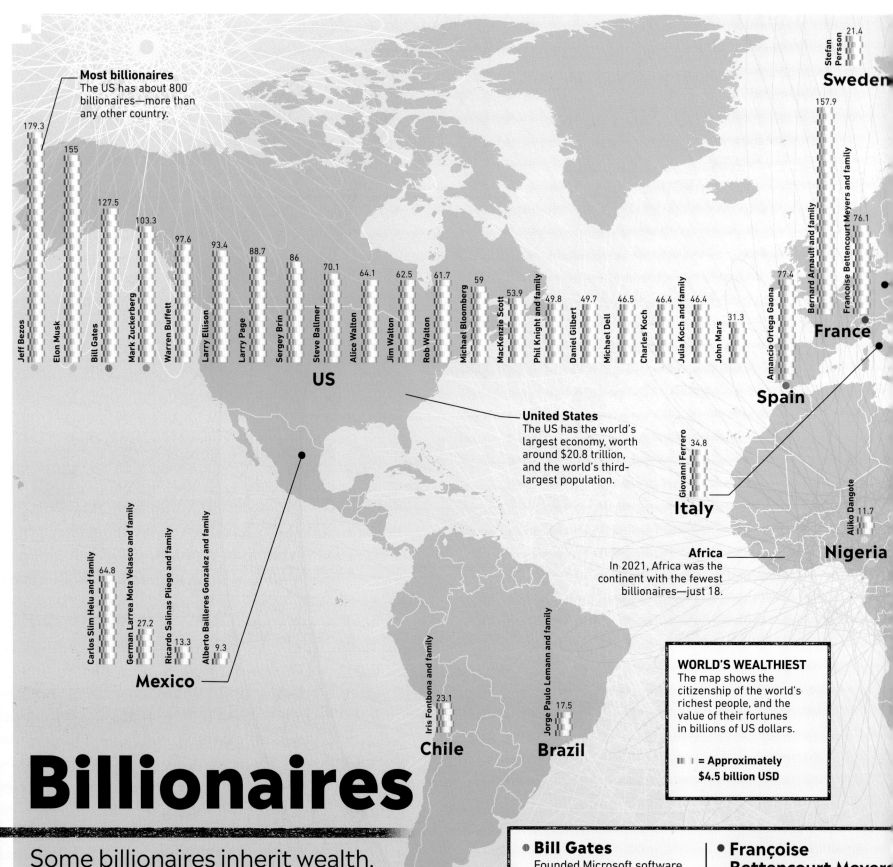

Billionaires

Some billionaires inherit wealth. Others get rich through banking, making or trading goods, or inventing new things. Not surprisingly, billionaires tend to be concentrated in more prosperous nations.

Most billionaires
The US has about 800 billionaires—more than any other country.

US

- Jeff Bezos — 179.3
- Elon Musk — 155
- Bill Gates — 127.5
- Mark Zuckerberg — 103.3
- Warren Buffett — 97.6
- Larry Ellison — 93.4
- Larry Page — 88.7
- Sergey Brin — 86
- Steve Ballmer — 70.1
- Alice Walton — 64.1
- Jim Walton — 62.5
- Rob Walton — 61.7
- Michael Bloomberg — 59
- MacKenzie Scott — 53.9
- Phil Knight and family — 49.8
- Daniel Gilbert — 49.7
- Michael Dell — 46.5
- Charles Koch — 46.4
- Julia Koch and family — 46.4
- John Mars — 31.3

United States
The US has the world's largest economy, worth around $20.8 trillion, and the world's third-largest population.

Mexico

- Carlos Slim Helu and family — 64.8
- German Larrea Mota Velasco and family — 27.2
- Ricardo Salinas Pliego and family — 13.3
- Alberto Bailleres Gonzalez and family — 9.3

Chile
- Iris Fontbona and family — 23.1

Brazil
- Jorge Paulo Lemann and family — 17.5

Sweden
- Stefan Persson — 21.4

France
- Bernard Arnault and family — 157.9
- Françoise Bettencourt Meyers and family — 76.1

Spain
- Amancio Ortega Gaona — 77.4

Italy
- Giovanni Ferrero — 34.8

Africa
In 2021, Africa was the continent with the fewest billionaires—just 18.

Nigeria
- Aliko Dangote — 11.7

WORLD'S WEALTHIEST
The map shows the citizenship of the world's richest people, and the value of their fortunes in billions of US dollars.

▌▌ ▌ = Approximately $4.5 billion USD

● **Bill Gates**
Founded Microsoft software firm in 1975. Now devotes himself to charity work.

● **Françoise Bettencourt Meyers**
A principal shareholder in the beauty company L'Oréal.

SINCE 1994, THE BILL AND MELINDA GATES FOUNDATION HAS GIVEN

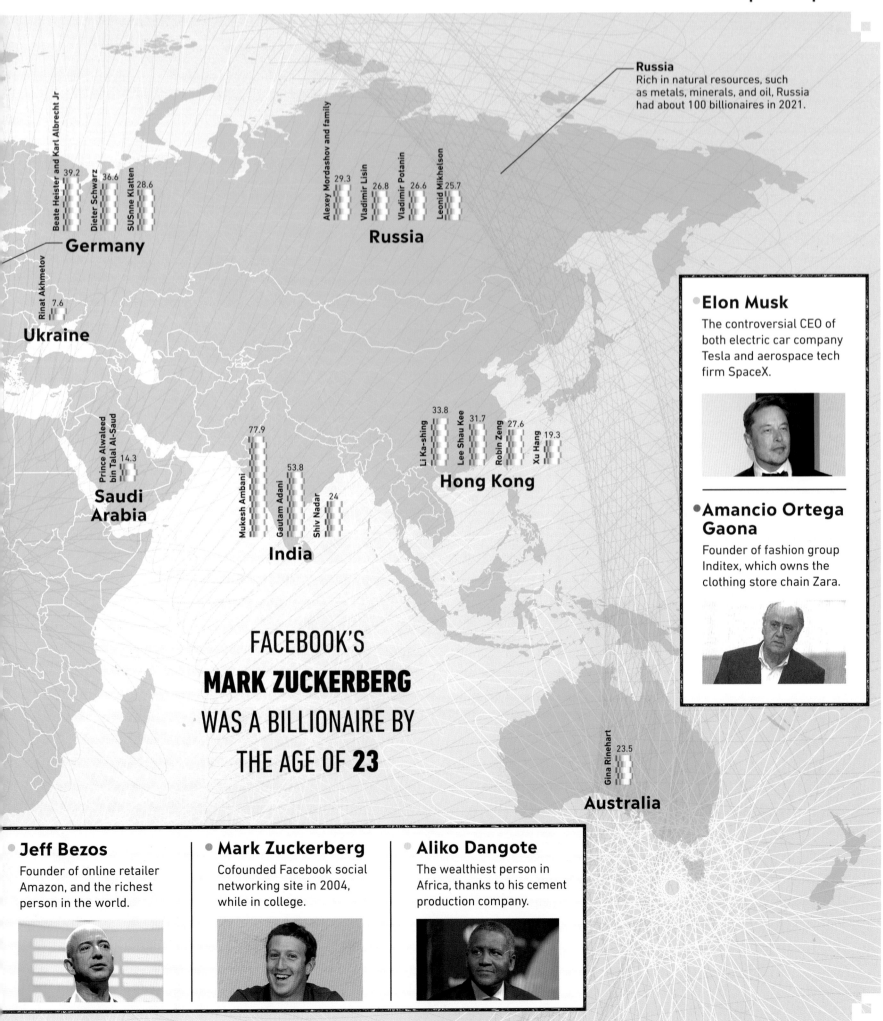

Russia
Rich in natural resources, such as metals, minerals, and oil, Russia had about 100 billionaires in 2021.

Germany

- Beate Heister and Karl Albrecht Jr — 39.2
- Dieter Schwarz — 36.6
- SUSnne Klatten — 28.6

Russia

- Alexey Mordashov and family — 29.3
- Vladimir Lisin — 26.8
- Vladimir Potanin — 26.6
- Leonid Mikhelson — 25.7

Ukraine

- Rinat Akhmetov — 7.6

Saudi Arabia

- Prince Alwaleed bin Talal Al-Saud — 14.3

India

- Mukesh Ambani — 77.9
- Gautam Adani — 53.8
- Shiv Nadar — 24

Hong Kong

- Li Ka-shing — 33.8
- Lee Shau Kee — 31.7
- Robin Zeng — 27.6
- Xu Hang — 19.3

Australia

- Gina Rinehart — 23.5

FACEBOOK'S

MARK ZUCKERBERG

WAS A BILLIONAIRE BY

THE AGE OF **23**

•Elon Musk
The controversial CEO of both electric car company Tesla and aerospace tech firm SpaceX.

•Amancio Ortega Gaona
Founder of fashion group Inditex, which owns the clothing store chain Zara.

•Jeff Bezos
Founder of online retailer Amazon, and the richest person in the world.

•Mark Zuckerberg
Cofounded Facebook social networking site in 2004, while in college.

•Aliko Dangote
The wealthiest person in Africa, thanks to his cement production company.

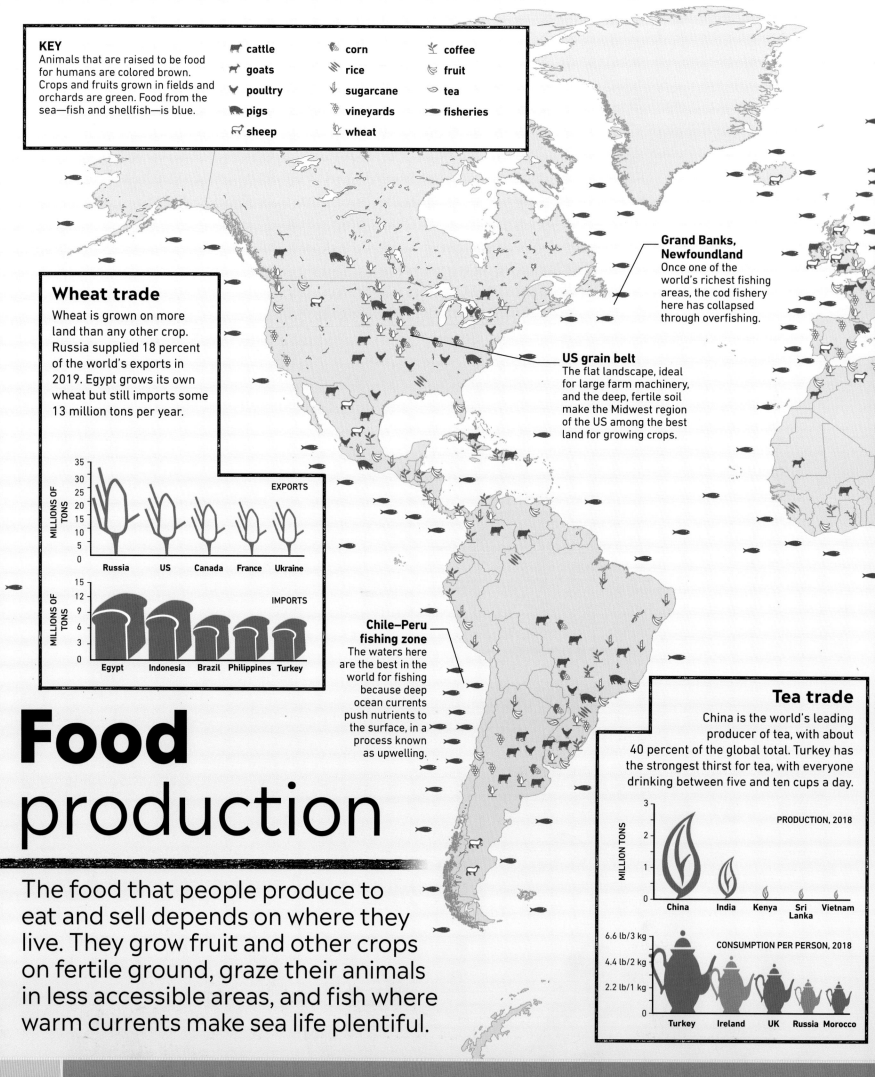

KEY

Animals that are raised to be food for humans are colored brown. Crops and fruits grown in fields and orchards are green. Food from the sea—fish and shellfish—is blue.

- 🐂 cattle
- 🐐 goats
- 🦃 poultry
- 🐖 pigs
- 🐑 sheep
- 🌽 corn
- 🌾 rice
- 🌱 sugarcane
- 🍇 vineyards
- 🌾 wheat
- ☘ coffee
- 🍌 fruit
- 🍃 tea
- 🐟 fisheries

Wheat trade

Wheat is grown on more land than any other crop. Russia supplied 18 percent of the world's exports in 2019. Egypt grows its own wheat but still imports some 13 million tons per year.

EXPORTS

MILLIONS OF TONS

35 30 25 20 15 10 5 0

Russia | US | Canada | France | Ukraine

IMPORTS

MILLIONS OF TONS

15 12 9 6 3 0

Egypt | Indonesia | Brazil | Philippines | Turkey

Grand Banks, Newfoundland

Once one of the world's richest fishing areas, the cod fishery here has collapsed through overfishing.

US grain belt

The flat landscape, ideal for large farm machinery, and the deep, fertile soil make the Midwest region of the US among the best land for growing crops.

Chile–Peru fishing zone

The waters here are the best in the world for fishing because deep ocean currents push nutrients to the surface, in a process known as upwelling.

Food production

The food that people produce to eat and sell depends on where they live. They grow fruit and other crops on fertile ground, graze their animals in less accessible areas, and fish where warm currents make sea life plentiful.

Tea trade

China is the world's leading producer of tea, with about 40 percent of the global total. Turkey has the strongest thirst for tea, with everyone drinking between five and ten cups a day.

PRODUCTION, 2018

MILLION TONS

3 2 1 0

China | India | Kenya | Sri Lanka | Vietnam

CONSUMPTION PER PERSON, 2018

6.6 lb/3 kg
4.4 lb/2 kg
2.2 lb/1 kg
0

Turkey | Ireland | UK | Russia | Morocco

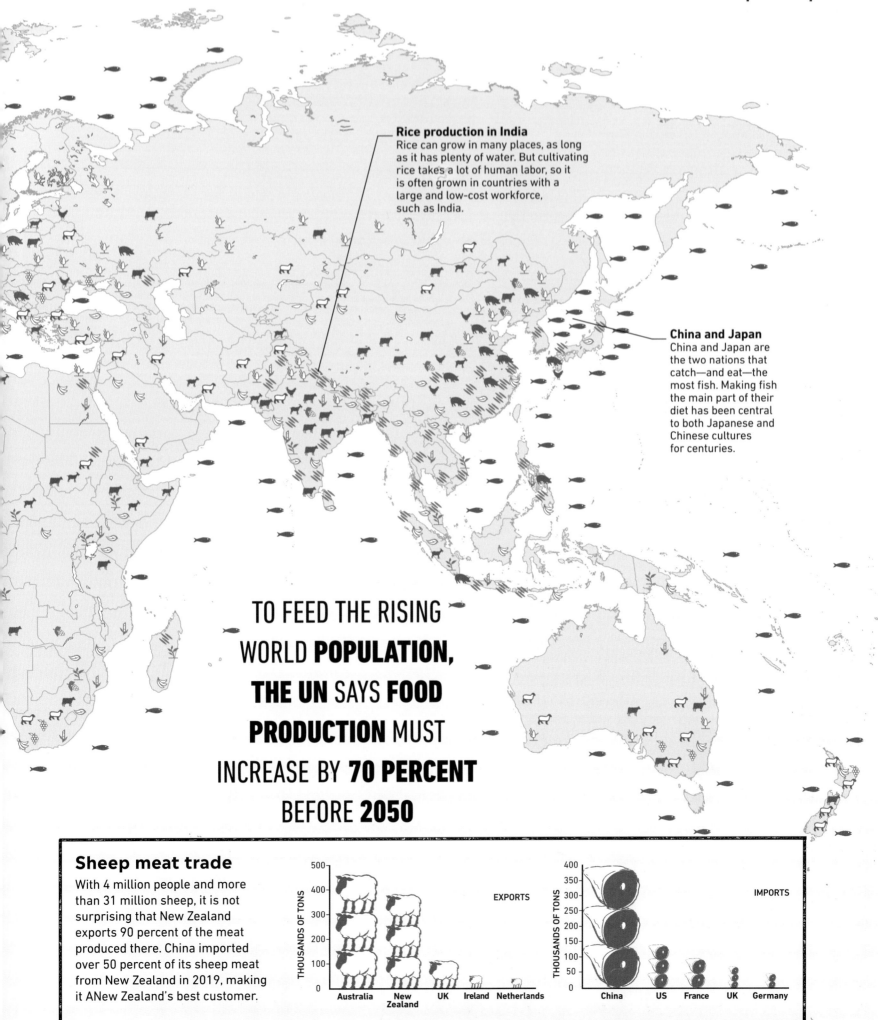

Rice production in India
Rice can grow in many places, as long as it has plenty of water. But cultivating rice takes a lot of human labor, so it is often grown in countries with a large and low-cost workforce, such as India.

China and Japan
China and Japan are the two nations that catch—and eat—the most fish. Making fish the main part of their diet has been central to both Japanese and Chinese cultures for centuries.

TO FEED THE RISING WORLD **POPULATION, THE UN** SAYS **FOOD PRODUCTION** MUST INCREASE BY **70 PERCENT** BEFORE **2050**

Sheep meat trade

With 4 million people and more than 31 million sheep, it is not surprising that New Zealand exports 90 percent of the meat produced there. China imported over 50 percent of its sheep meat from New Zealand in 2019, making it ANew Zealand's best customer.

EXPORTS

THOUSANDS OF TONS
500
400
300
200
100
0
Australia | New Zealand | UK | Ireland | Netherlands

IMPORTS

THOUSANDS OF TONS
400
350
300
250
200
150
100
50
0
China | US | France | UK | Germany

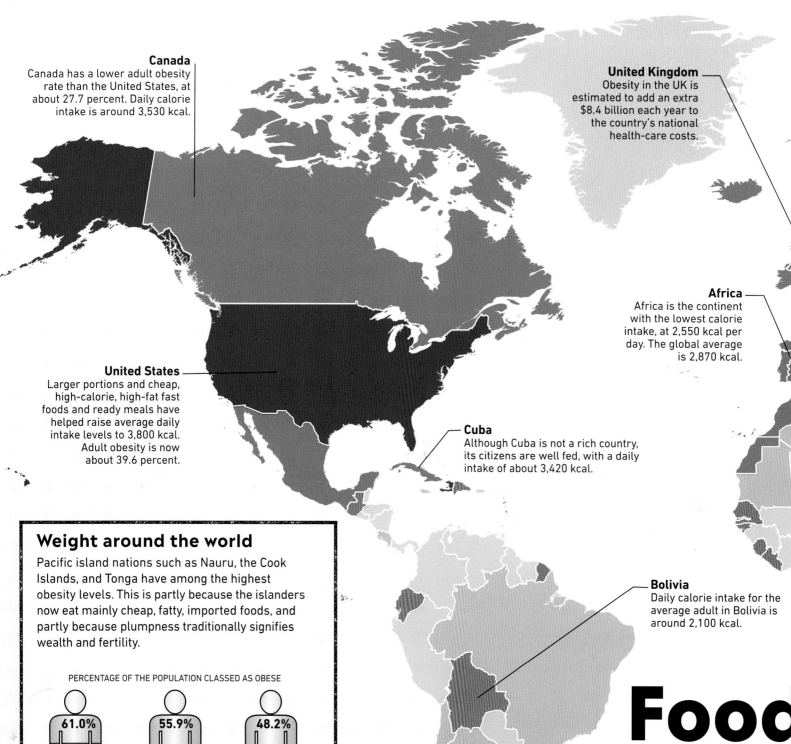

Canada
Canada has a lower adult obesity rate than the United States, at about 27.7 percent. Daily calorie intake is around 3,530 kcal.

United Kingdom
Obesity in the UK is estimated to add an extra $8.4 billion each year to the country's national health-care costs.

Africa
Africa is the continent with the lowest calorie intake, at 2,550 kcal per day. The global average is 2,870 kcal.

United States
Larger portions and cheap, high-calorie, high-fat fast foods and ready meals have helped raise average daily intake levels to 3,800 kcal. Adult obesity is now about 39.6 percent.

Cuba
Although Cuba is not a rich country, its citizens are well fed, with a daily intake of about 3,420 kcal.

Bolivia
Daily calorie intake for the average adult in Bolivia is around 2,100 kcal.

Weight around the world
Pacific island nations such as Nauru, the Cook Islands, and Tonga have among the highest obesity levels. This is partly because the islanders now eat mainly cheap, fatty, imported foods, and partly because plumpness traditionally signifies wealth and fertility.

PERCENTAGE OF THE POPULATION CLASSED AS OBESE

61.0%	55.9%	48.2%
Nauru	**Cook Islands**	**Tonga**
36.2%	35.4%	31.7%
US	**Saudi Arabia**	**United Arab Emirates**

Food intake

Food and the energy it contains is the fuel for our bodies. Overeating and unhealthy diets can lead to obesity—when a person gains so much weight that it can cause illness and disease.

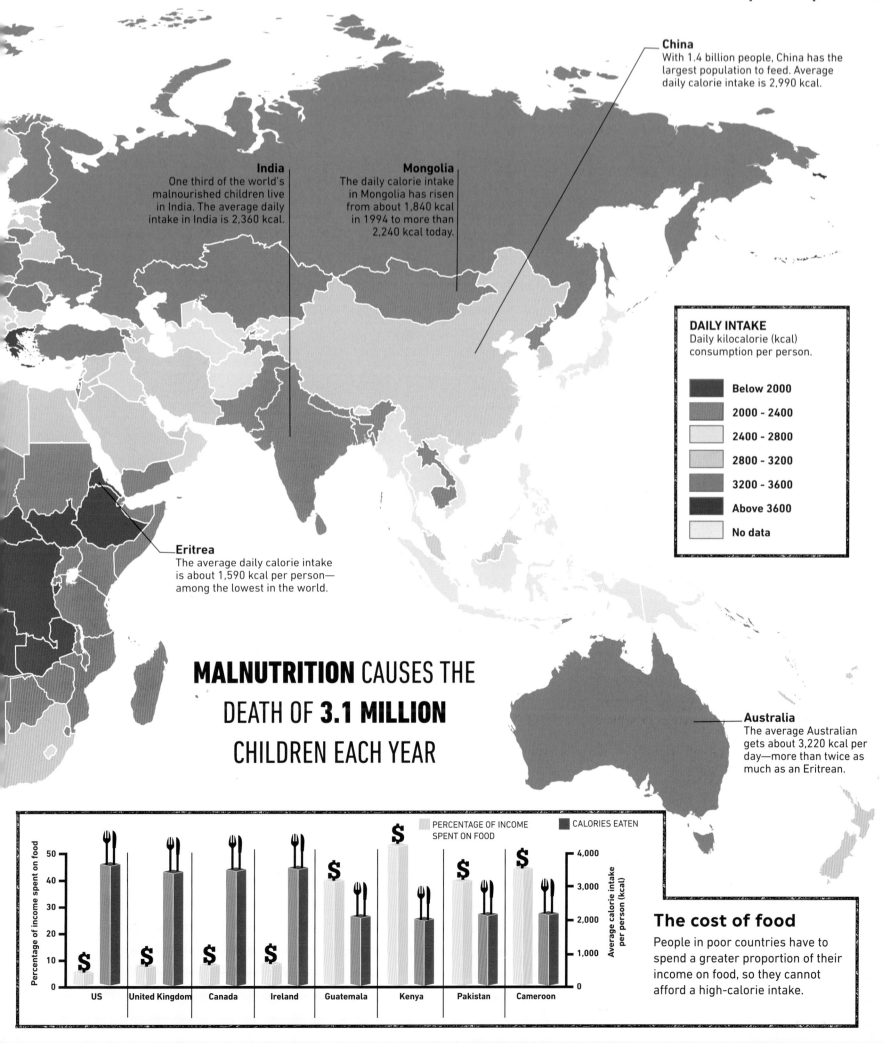

China
With 1.4 billion people, China has the largest population to feed. Average daily calorie intake is 2,990 kcal.

India
One third of the world's malnourished children live in India. The average daily intake in India is 2,360 kcal.

Mongolia
The daily calorie intake in Mongolia has risen from about 1,840 kcal in 1994 to more than 2,240 kcal today.

DAILY INTAKE
Daily kilocalorie (kcal) consumption per person.

- Below 2000
- 2000 – 2400
- 2400 – 2800
- 2800 – 3200
- 3200 – 3600
- Above 3600
- No data

Eritrea
The average daily calorie intake is about 1,590 kcal per person—among the lowest in the world.

MALNUTRITION CAUSES THE DEATH OF **3.1 MILLION** CHILDREN EACH YEAR

Australia
The average Australian gets about 3,220 kcal per day—more than twice as much as an Eritrean.

PERCENTAGE OF INCOME SPENT ON FOOD   CALORIES EATEN

Percentage of income spent on food

Average calorie intake per person (kcal)

US   United Kingdom   Canada   Ireland   Guatemala   Kenya   Pakistan   Cameroon

The cost of food

People in poor countries have to spend a greater proportion of their income on food, so they cannot afford a high-calorie intake.

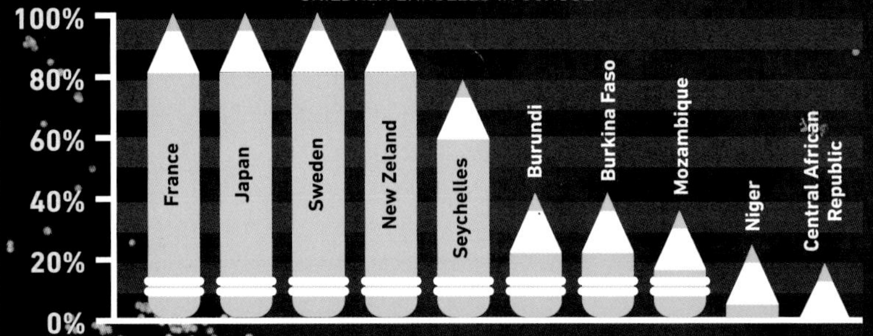

Canada
About 16 percent of Canadians struggle to pass basic literary tests.

Europe
Although most countries in Europe have very high literacy rates, more than 55 million adults classed as "literate" still lack basic reading and writing skills.

United States
About 21 percent of adults in the US are classed as "functionally illiterate."

Mauritania
Little more than half of Mauritania's population—52.1 percent—can read and write.

Chad
Just 23.3 percent of people in Chad are literate—the world's lowest literacy rate.

Brazil
Just over nine out of every ten Brazilians are literate.

Going to secondary school

Wealthy nations can afford to provide secondary education for all children, but governments in poorer countries cannot offer every child a place. This is particularly true in Africa south of the Sahara. In Niger, for example, only 24 percent of children go to secondary school.

PERCENTAGE OF SECONDARY-SCHOOL-AGE CHILDREN ENROLLED IN SCHOOL

100%	
80%	France, Japan, Sweden, New Zeland
60%	Seychelles
40%	Burundi, Burkina Faso, Mozambique
20%	Niger, Central African Republic
0%	

Literacy

Literacy—being able to read and write—is an essential life skill. Being literate makes it easier for people to learn, make the most of their abilities, and get better jobs. High levels of illiteracy make it difficult for nations to develop and become wealthier.

IN DEVELOPING COUNTRIES, **200 MILLION** PEOPLE AGES 15–24 HAVE NOT COMPLETED **PRIMARY SCHOOL**

ABOUT 61 MILLION PRIMARY-AGE CHILDREN WORLDWIDE WERE RECEIVING

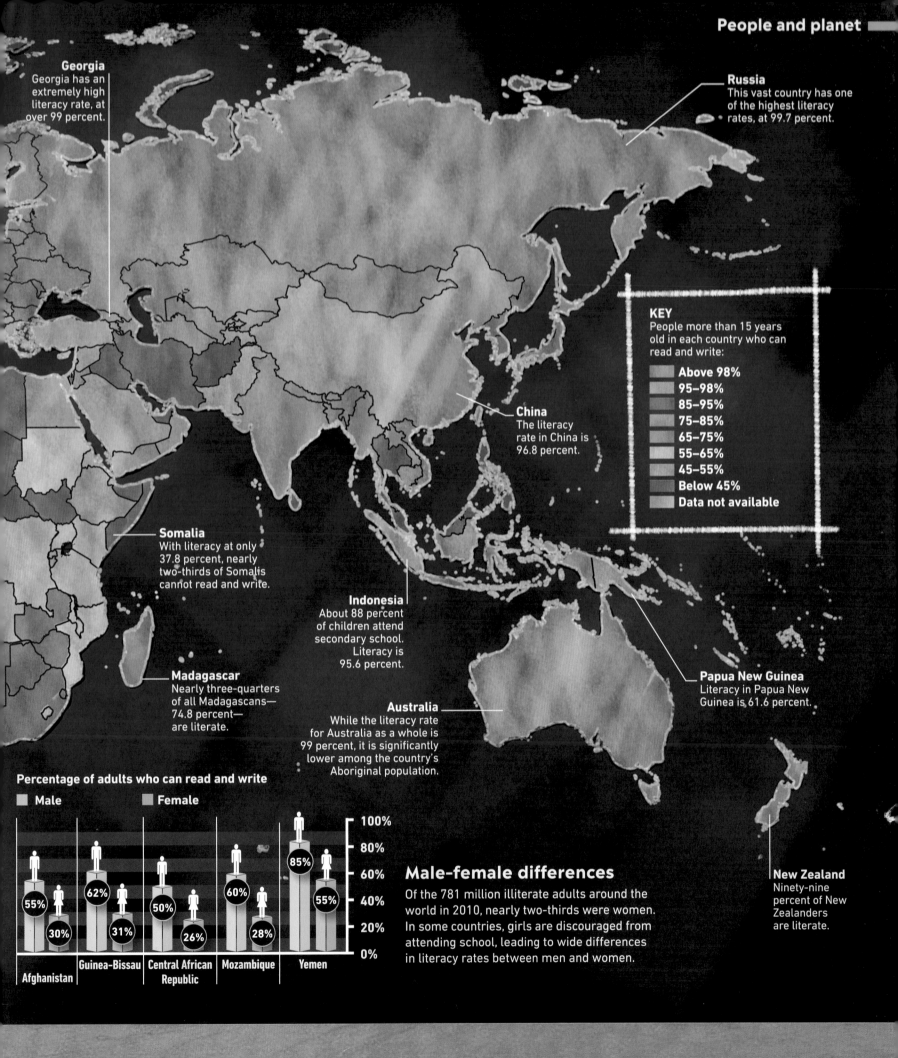

Georgia
Georgia has an extremely high literacy rate, at over 99 percent.

Russia
This vast country has one of the highest literacy rates, at 99.7 percent.

China
The literacy rate in China is 96.8 percent.

Somalia
With literacy at only 37.8 percent, nearly two-thirds of Somalis cannot read and write.

Indonesia
About 88 percent of children attend secondary school. Literacy is 95.6 percent.

Madagascar
Nearly three-quarters of all Madagascans—74.8 percent—are literate.

Australia
While the literacy rate for Australia as a whole is 99 percent, it is significantly lower among the country's Aboriginal population.

Papua New Guinea
Literacy in Papua New Guinea is 61.6 percent.

New Zealand
Ninety-nine percent of New Zealanders are literate.

KEY
People more than 15 years old in each country who can read and write:

- Above 98%
- 95–98%
- 85–95%
- 75–85%
- 65–75%
- 55–65%
- 45–55%
- Below 45%
- Data not available

Percentage of adults who can read and write

■ Male   ■ Female

	Afghanistan	Guinea-Bissau	Central African Republic	Mozambique	Yemen
Male	55%	62%	50%	60%	85%
Female	30%	31%	26%	28%	55%

100%
80%
60%
40%
20%
0%

Male-female differences
Of the 781 million illiterate adults around the world in 2010, nearly two-thirds were women. In some countries, girls are discouraged from attending school, leading to wide differences in literacy rates between men and women.

Biggest oil spills

Oil spills—when oil escapes into the environment—cause devastation to wildlife and are difficult and costly to clean up.

1 Gulf War oil spill, Persian Gulf, 1991
330,000–1,322,000 tons
Iraqi forces opened valves on Kuwaiti oil wells and pipes, causing a 100-mile (160-km) slick.

2 Lakeview gusher, California 1910-11
1,212,000 tons
An oil well erupted like a geyser, spilling out oil for over a year until it naturally died down.

3 Deepwater Horizon, Gulf of Mexico, 2010
740,000 tons
A deep-sea oil spill occurred when an explosion destroyed the Deepwater Horizon drillling rig.

4 Ixtoc 1 oil spill, Gulf of Mexico, USA, 1979–80
454,000–480,000 tons
The Ixtoc 1 drilling platform collapsed after an explosion. The spill continued for 9 months.

5 Atlantic Empress, Trinidad and Tobago, 1979
287,000 tons
The largest oil spill from a ship. The tanker *Atlantic Empress* hit another ship, killing 26 crew.

Pollution

Oil spills, industrial waste, and radiation leaks from nuclear power stations cause harm to people and the environment. Carbon dioxide gas (CO_2) produced by transportation and industry is adding to global warming.

Persistent organic pollutants (POPs): Canadian Arctic
These pollutants include industrial products and pesticides. They travel on the world's oceans and air currents, accumulate in the Arctic regions, and contaminate the foods that Inuit people eat.

Lead: La Oroya, Peru
A metal smelting plant has emitted toxic lead since 1922. This has led to contaminated water supplies, dangerously polluted air, and unsafe levels of lead in the blood of local residents.

THE 1991 **GULF WAR** OIL SLICK WAS UP TO **5 IN (13 CM) THICK**

Nuclear accidents
Splitting atoms in nuclear reactors produces energy for generating electricity. Accidents at reactors may lead to radioactive material escaping, which can cause illness such as cancer for many years.

 1 Chernobyl, Ukraine April 26, 1986
A reactor explosion released radioactive material. Radiation-related illnesses may have caused thousands of deaths.

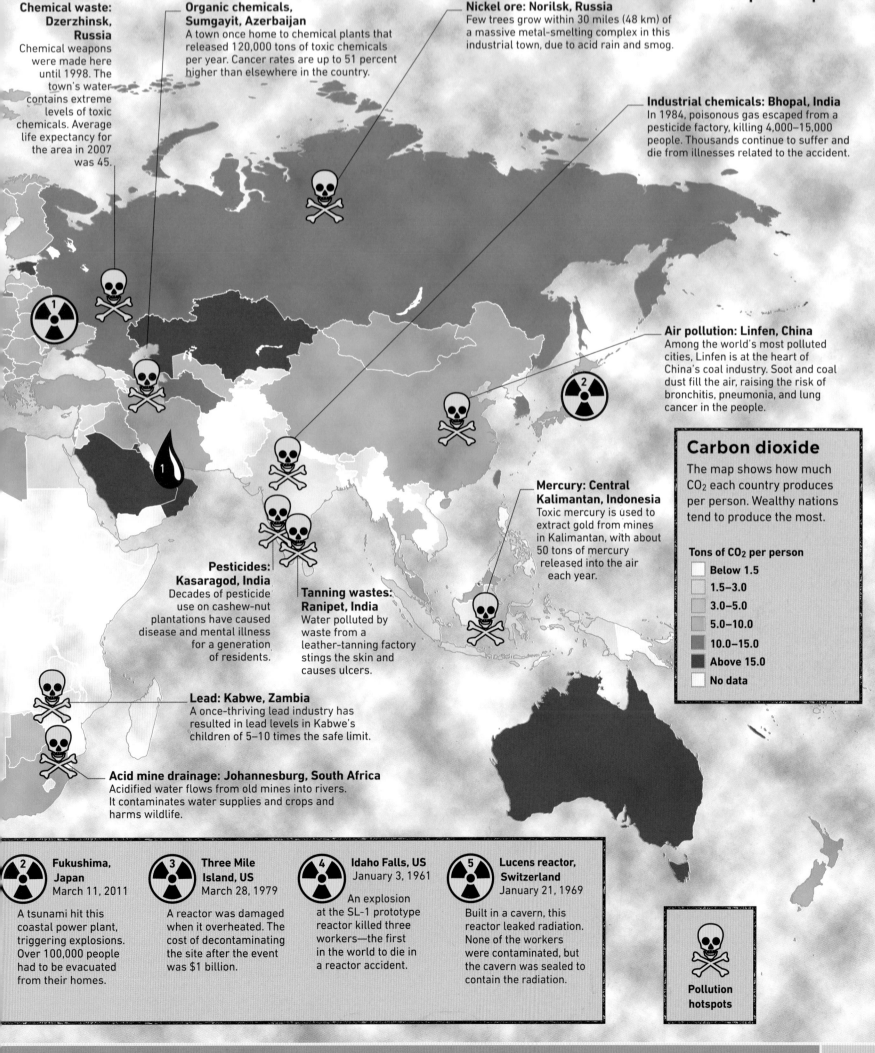

Chemical waste: Dzerzhinsk, Russia
Chemical weapons were made here until 1998. The town's water contains extreme levels of toxic chemicals. Average life expectancy for the area in 2007 was 45.

Organic chemicals, Sumgayit, Azerbaijan
A town once home to chemical plants that released 120,000 tons of toxic chemicals per year. Cancer rates are up to 51 percent higher than elsewhere in the country.

Nickel ore: Norilsk, Russia
Few trees grow within 30 miles (48 km) of a massive metal-smelting complex in this industrial town, due to acid rain and smog.

Industrial chemicals: Bhopal, India
In 1984, poisonous gas escaped from a pesticide factory, killing 4,000–15,000 people. Thousands continue to suffer and die from illnesses related to the accident.

Air pollution: Linfen, China
Among the world's most polluted cities, Linfen is at the heart of China's coal industry. Soot and coal dust fill the air, raising the risk of bronchitis, pneumonia, and lung cancer in the people.

Mercury: Central Kalimantan, Indonesia
Toxic mercury is used to extract gold from mines in Kalimantan, with about 50 tons of mercury released into the air each year.

Pesticides: Kasaragod, India
Decades of pesticide use on cashew-nut plantations have caused disease and mental illness for a generation of residents.

Tanning wastes: Ranipet, India
Water polluted by waste from a leather-tanning factory stings the skin and causes ulcers.

Lead: Kabwe, Zambia
A once-thriving lead industry has resulted in lead levels in Kabwe's children of 5–10 times the safe limit.

Acid mine drainage: Johannesburg, South Africa
Acidified water flows from old mines into rivers. It contaminates water supplies and crops and harms wildlife.

Carbon dioxide

The map shows how much CO_2 each country produces per person. Wealthy nations tend to produce the most.

Tons of CO_2 per person

- Below 1.5
- 1.5–3.0
- 3.0–5.0
- 5.0–10.0
- 10.0–15.0
- Above 15.0
- No data

2 Fukushima, Japan
March 11, 2011
A tsunami hit this coastal power plant, triggering explosions. Over 100,000 people had to be evacuated from their homes.

3 Three Mile Island, US
March 28, 1979
A reactor was damaged when it overheated. The cost of decontaminating the site after the event was $1 billion.

4 Idaho Falls, US
January 3, 1961
An explosion at the SL-1 prototype reactor killed three workers—the first in the world to die in a reactor accident.

5 Lucens reactor, Switzerland
January 21, 1969
Built in a cavern, this reactor leaked radiation. None of the workers were contaminated, but the cavern was sealed to contain the radiation.

Pollution hotspots

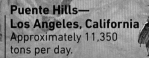

Puente Hills—Los Angeles, California
Approximately 11,350 tons per day.

Apex—Las Vegas, Nevada
Approximately 11,600 tons per day.

Greenland
Currently Greenland produces 30% more waste than it can process, though two new garbage-to-energy incinerators are due to open in 2021 and 2022.

Western Pacific Garbage Patch
A lot of discarded litter ends up in rivers, which take it to the sea, where circular currents called gyres collect it into vast patches in the ocean surface waters. This patch is the largest of these oceanic rubbish dumps.

North Atlantic Garbage Patch
The North Atlantic Garbage Patch measures hundreds of miles across. It shifts by as much as 990 miles (1,600 km) north and south with the seasons.

Bordo Poniente Landfill—Nezahualcoyotl, Mexico
Over 13,200 tons per day.

Gabon
Less wealthy countries, such as Gabon, produce less garbage because people buy less overall, they buy proportionally more local produce without plastic packaging, and do more recycling.

Garbage and waste

As living standards improve worldwide and cities grow, so does the amount of garbage that people produce. Most waste goes to garbage dumps, which are expensive, use up a lot of land, and are harmful to the environment. Recycling is one way of helping to stop the global garbage heap from growing any bigger.

South Pacific Garbage Patch
So far, the South Pacific Gyre appears to contain less plastic waste than other ocean garbage patches.

South Atlantic Garbage Patch
The first evidence of a South Atlantic Garbage Patch was discovered in 2011. Most plastic particles in ocean garbage patches are too small to be seen with the naked eye.

Top of the recycling table
Only a handful of countries currently recycle more than half their waste; Germany tops this list, recycling 56.1% of all waste in 2019. This figure is a rapid increase from 1991, when the country recycled only 3% of its garbage.

RECYCLING ONE ALUMINIUM CAN SAVES ENOUGH ENERGY TO BURN A

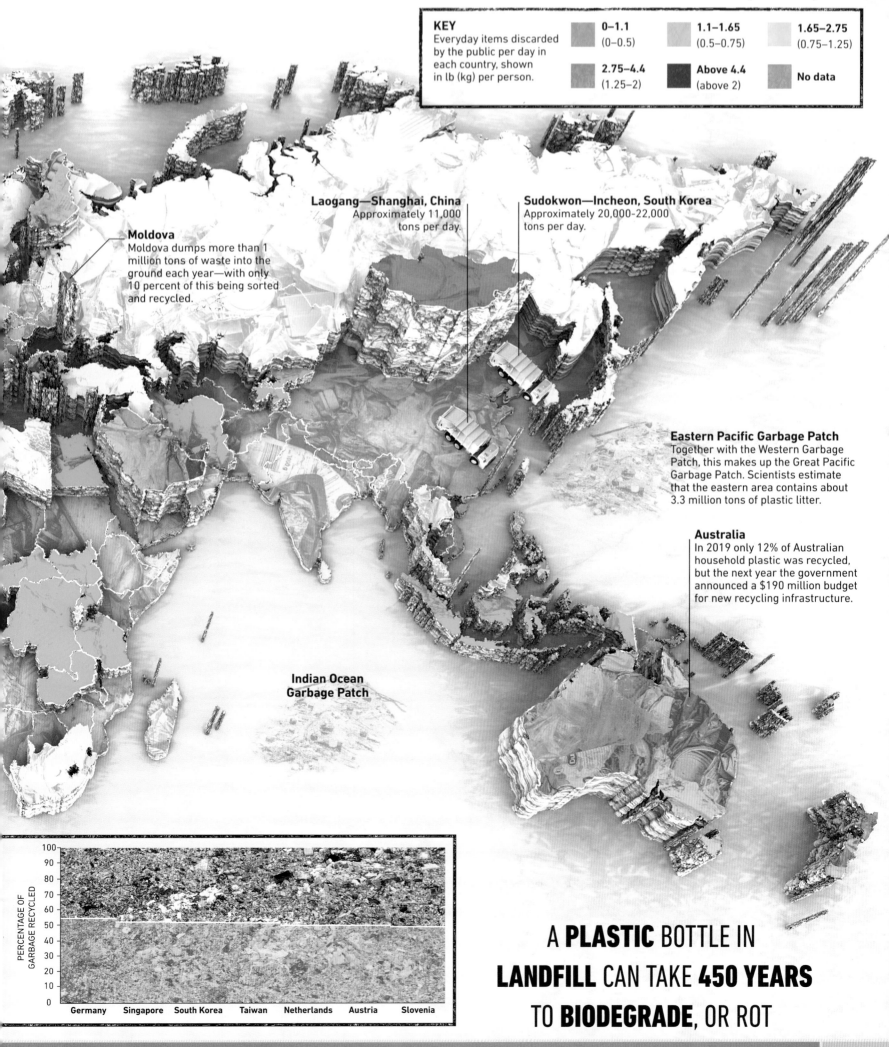

KEY
Everyday items discarded by the public per day in each country, shown in lb (kg) per person.

0–1.1 (0–0.5)	
1.1–1.65 (0.5–0.75)	
1.65–2.75 (0.75–1.25)	
2.75–4.4 (1.25–2)	
Above 4.4 (above 2)	
No data	

Laogang—Shanghai, China
Approximately 11,000 tons per day.

Sudokwon—Incheon, South Korea
Approximately 20,000–22,000 tons per day.

Moldova
Moldova dumps more than 1 million tons of waste into the ground each year—with only 10 percent of this being sorted and recycled.

Eastern Pacific Garbage Patch
Together with the Western Garbage Patch, this makes up the Great Pacific Garbage Patch. Scientists estimate that the eastern area contains about 3.3 million tons of plastic litter.

Australia
In 2019 only 12% of Australian household plastic was recycled, but the next year the government announced a $190 million budget for new recycling infrastructure.

Indian Ocean Garbage Patch

PERCENTAGE OF GARBAGE RECYCLED

100 90 80 70 60 50 40 30 20 10 0

Germany Singapore South Korea Taiwan Netherlands Austria Slovenia

A **PLASTIC** BOTTLE IN **LANDFILL** CAN TAKE **450 YEARS** TO **BIODEGRADE**, OR ROT

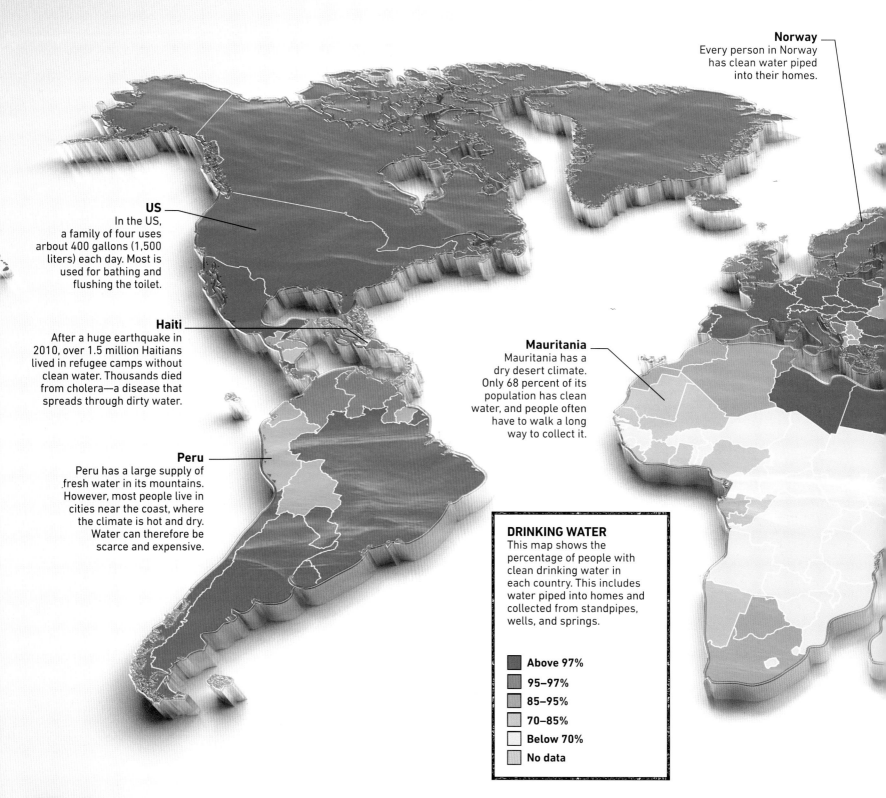

Norway
Every person in Norway has clean water piped into their homes.

US
In the US, a family of four uses arbout 400 gallons (1,500 liters) each day. Most is used for bathing and flushing the toilet.

Haiti
After a huge earthquake in 2010, over 1.5 million Haitians lived in refugee camps without clean water. Thousands died from cholera—a disease that spreads through dirty water.

Mauritania
Mauritania has a dry desert climate. Only 68 percent of its population has clean water, and people often have to walk a long way to collect it.

Peru
Peru has a large supply of fresh water in its mountains. However, most people live in cities near the coast, where the climate is hot and dry. Water can therefore be scarce and expensive.

DRINKING WATER
This map shows the percentage of people with clean drinking water in each country. This includes water piped into homes and collected from standpipes, wells, and springs.

- Above 97%
- 95–97%
- 85–95%
- 70–85%
- Below 70%
- No data

Clean water

The tap in your home may give you an instant supply of clean drinking water. However, millions of people around the world must get their water from a standpipe or a well. For one in three people, their sources of water are contaminated and unsafe to drink.

Thirsty crops

Growing crops in dry climates is by far the thirstiest human activity. It uses much more water than is used in people's homes and dominates water use in many countries. That's why parts of central Asia, where farmers water fields of cotton, top this list of overall water consumers.

IN DEVELOPING COUNTRIES, 70 PERCENT OF INDUSTRIAL WASTE IS

Russia
Russia's rivers and lakes provide plentiful water, but the quality of water supplies is not reliable and most people must buy bottled water to drink.

Water use in the home
These glasses show how many of gallons (liters) of water each person uses a day for such things as drinking, washing, cooking, and cleaning. In Cambodia, each person manages on just 5 gallons (19 liters) of water a day. In the US, people use 30 times more.

Cambodia 5.2 (19.6)
Kenya 9.1 (34.4)
Brazil 61.9 (234)
Kuwait 133 (501)
USA 157 (593)

India
Nineteen percent of the world's population without clean water access live in India. About 850 children here under the age of five die every day from diarrhea.

Indonesia
Many water supplies are polluted by waste from factories and by sewage. About 70 percent of people lack clean water.

THE UN SAYS EACH PERSON SHOULD HAVE 13 GALLONS (50 LITERS) OF CLEAN WATER EVERY DAY

Total water use (million liters per person per year)

0 1 2 3 4 5 6

- Turkmenistan
- Chile
- Guyana
- Uzbekistan
- Tajikistan
- Kyrgyzstan
- US
- Iran
- Estonia
- Azerbaijan

Australia
A history of terrible droughts caused Australia rethink its water use. Measures include recycling sewage and encouraging gray water recycling (waste water from baths and washing machines), in an attempt to "drought-proof" the nation.

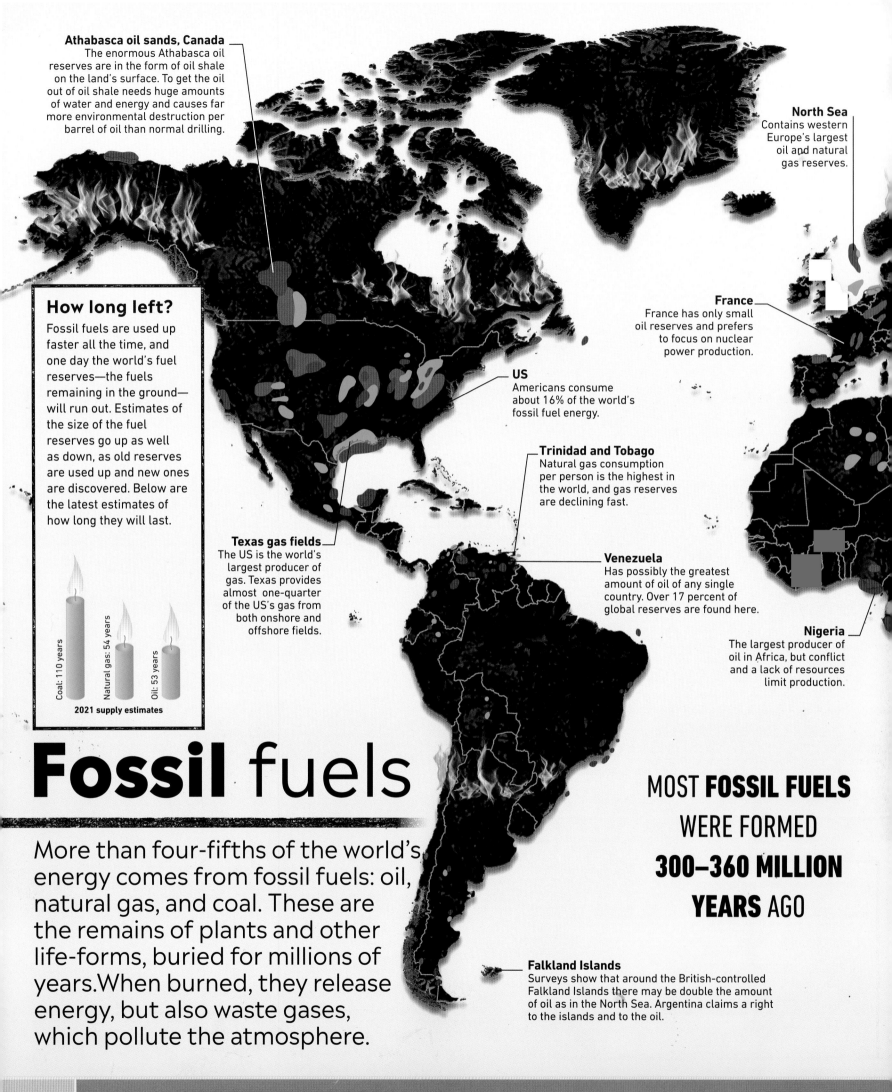

Athabasca oil sands, Canada
The enormous Athabasca oil reserves are in the form of oil shale on the land's surface. To get the oil out of oil shale needs huge amounts of water and energy and causes far more environmental destruction per barrel of oil than normal drilling.

North Sea
Contains western Europe's largest oil and natural gas reserves.

How long left?

Fossil fuels are used up faster all the time, and one day the world's fuel reserves—the fuels remaining in the ground—will run out. Estimates of the size of the fuel reserves go up as well as down, as old reserves are used up and new ones are discovered. Below are the latest estimates of how long they will last.

Coal: 110 years

Natural gas: 54 years

Oil: 53 years

2021 supply estimates

France
France has only small oil reserves and prefers to focus on nuclear power production.

US
Americans consume about 16% of the world's fossil fuel energy.

Trinidad and Tobago
Natural gas consumption per person is the highest in the world, and gas reserves are declining fast.

Texas gas fields
The US is the world's largest producer of gas. Texas provides almost one-quarter of the US's gas from both onshore and offshore fields.

Venezuela
Has possibly the greatest amount of oil of any single country. Over 17 percent of global reserves are found here.

Nigeria
The largest producer of oil in Africa, but conflict and a lack of resources limit production.

Fossil fuels

More than four-fifths of the world's energy comes from fossil fuels: oil, natural gas, and coal. These are the remains of plants and other life-forms, buried for millions of years. When burned, they release energy, but also waste gases, which pollute the atmosphere.

MOST FOSSIL FUELS WERE FORMED 300–360 MILLION YEARS AGO

Falkland Islands
Surveys show that around the British-controlled Falkland Islands there may be double the amount of oil as in the North Sea. Argentina claims a right to the islands and to the oil.

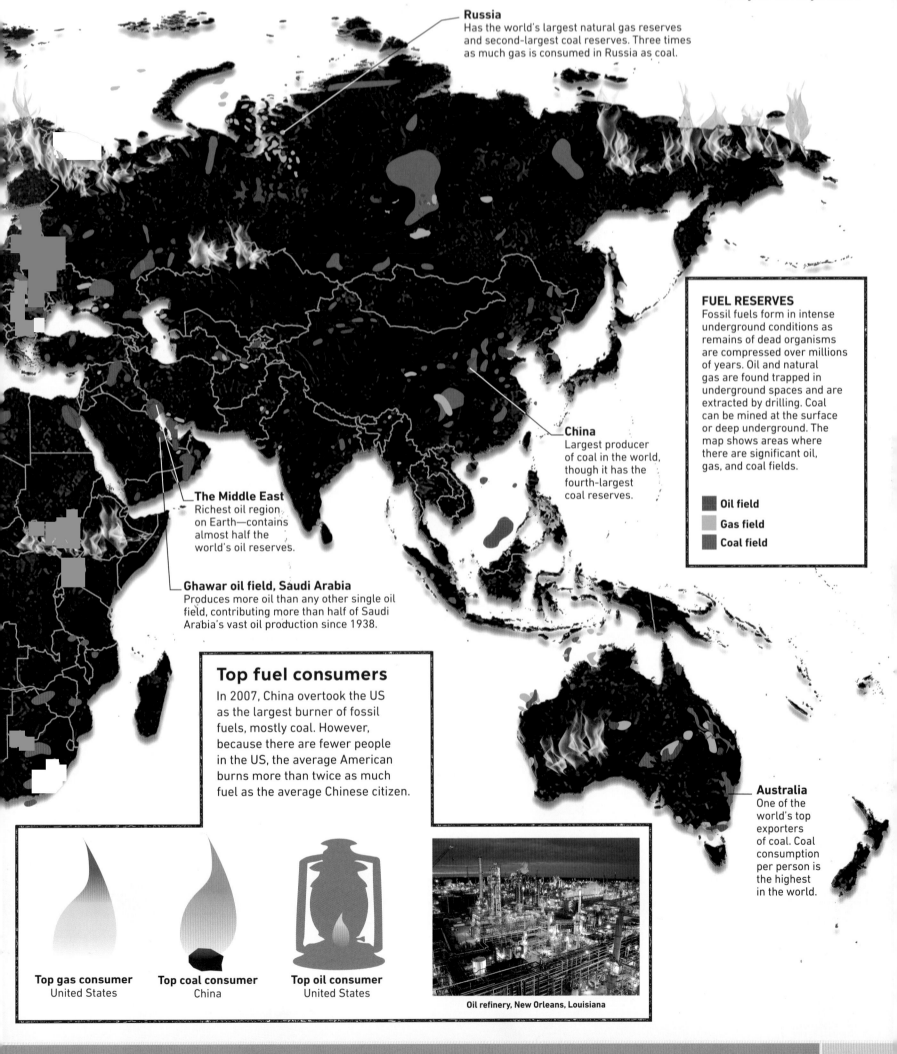

Russia
Has the world's largest natural gas reserves and second-largest coal reserves. Three times as much gas is consumed in Russia as coal.

China
Largest producer of coal in the world, though it has the fourth-largest coal reserves.

FUEL RESERVES
Fossil fuels form in intense underground conditions as remains of dead organisms are compressed over millions of years. Oil and natural gas are found trapped in underground spaces and are extracted by drilling. Coal can be mined at the surface or deep underground. The map shows areas where there are significant oil, gas, and coal fields.

- ■ Oil field
- ■ Gas field
- ■ Coal field

The Middle East
Richest oil region on Earth—contains almost half the world's oil reserves.

Ghawar oil field, Saudi Arabia
Produces more oil than any other single oil field, contributing more than half of Saudi Arabia's vast oil production since 1938.

Top fuel consumers

In 2007, China overtook the US as the largest burner of fossil fuels, mostly coal. However, because there are fewer people in the US, the average American burns more than twice as much fuel as the average Chinese citizen.

Australia
One of the world's top exporters of coal. Coal consumption per person is the highest in the world.

Top gas consumer
United States

Top coal consumer
China

Top oil consumer
United States

Oil refinery, New Orleans, Louisiana

COAL ON A LARGE SCALE TO HEAT HOUSES THROUGH UNDERFLOOR PIPES.

Alternative energy

There are several types of alternative energy, some of which are also renewable (see opposite page).

Wind
Mounted on tall masts, huge rotating blades called wind turbines harness the wind's energy and use it to drive electricity generators.

Solar
The sun's energy can be used to heat water in homes or to produce high temperatures for electricity generation. Photovoltaic panels convert sunlight directly into electricity.

Nuclear
The nuclei (cores) of atoms are split apart in nuclear power plants, releasing vast amounts of energy. However, the process also creates dangerous nuclear waste.

Geothermal
A geothermal power plant taps underground steam or hot water, which it uses to generate electricity or to heat buildings directly.

Hydroelectric
A hydroelectric power plant is a dam with generators built into it. Water builds up behind the dam. When gates in the dam are opened, the force of the falling water drives the generators.

Biofuel, biogas, and biomass
Liquid fuel made from plants, rather than oil, is called biofuel. When farm waste, sewage, and garbage rot, they release biogas, which can be burned as fuel. Biomass is any plant-based material burned for warmth or to generate electricity.

Alternative energy

Burning fossil fuels—coal, oil, and gas—creates a lot of pollution. People are developing alternative, cleaner energy sources, and some are renewable—they never run out.

Iceland
Ninth on the "top 10" list of geothermal producers

Norway
Number six for hydroelectricity

Germany
First in the world for solar, third for wind, fourth for biofuel, and eighth for nuclear

France
Ranked second for nuclear, fifth for biofuel, seventh for wind, and ninth for solar

Spain
Number five for wind and ten for biofuel

Italy
Ranked fourth for solar, seventh for geothermal, and tenth for wind

Canada
Fourth-largest producer of hydroelectricity; sixth for nuclear; ninth for wind

US
World's top producer of geothermal, biofuel, and nuclear energy; in second place for wind and hydroelectricity; and fifth for solar

Venezuela
Eighth for hydroelectricity

Mexico
Sixth for geothermal

El Salvador
Ninth for geothermal

Brazil
Ranked second for electricity production by biofuel, third for hydroelectricity, and eighth for wind

Argentina
Seventh-largest producer of electricity from biofuel

HYDROELECTRIC POWER PLANTS SUPPLY 90 PERCENT OF NORWAY'S ENERGY

KEY
Top 10 alternative-energy producing countries in each field

🌀 **Wind energy**

🔲 **Solar energy**

☢ **Nuclear energy**

⛰ **Geothermal energy**

≋ **Hydroelectric energy**

🌿 **Biofuel energy**

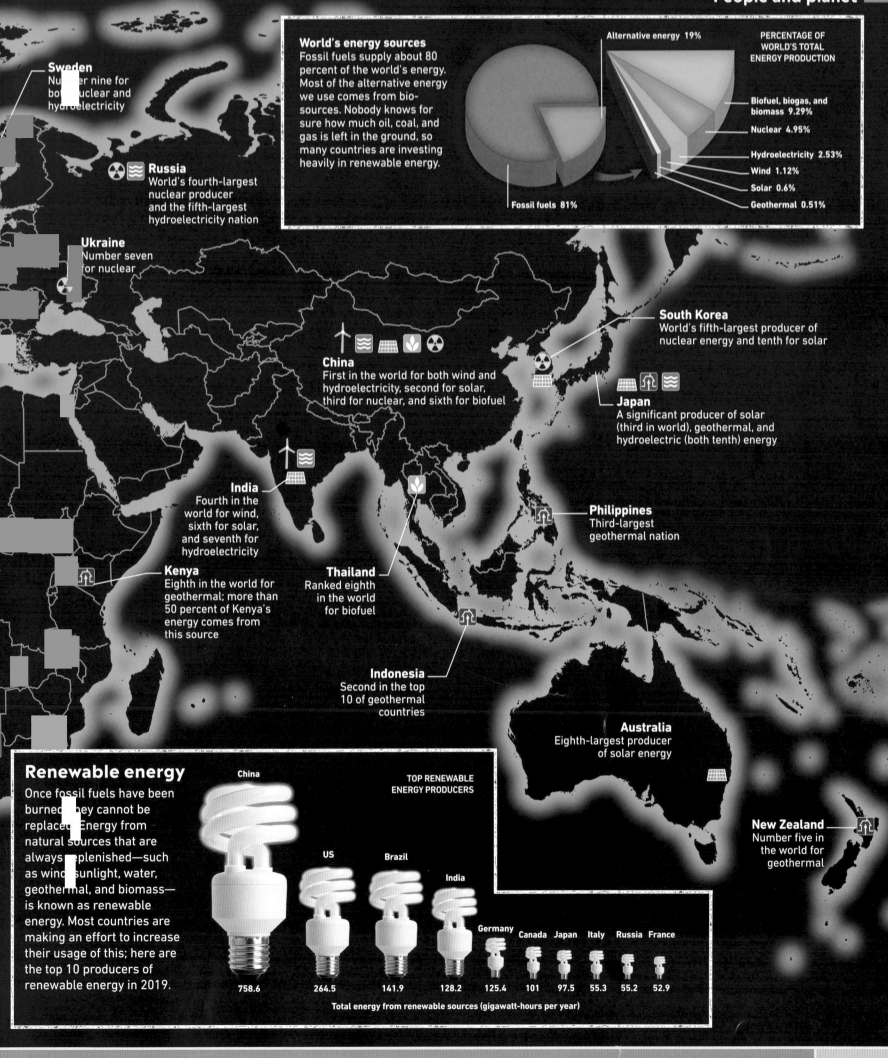

World's energy sources

Fossil fuels supply about 80 percent of the world's energy. Most of the alternative energy we use comes from bio-sources. Nobody knows for sure how much oil, coal, and gas is left in the ground, so many countries are investing heavily in renewable energy.

Alternative energy 19%

PERCENTAGE OF WORLD'S TOTAL ENERGY PRODUCTION

Biofuel, biogas, and biomass 9.29%

Nuclear 4.95%

Hydroelectricity 2.53%

Wind 1.12%

Solar 0.6%

Geothermal 0.51%

Fossil fuels 81%

Sweden
Number nine for both nuclear and hydroelectricity

Russia
World's fourth-largest nuclear producer and the fifth-largest hydroelectricity nation

Ukraine
Number seven for nuclear

China
First in the world for both wind and hydroelectricity, second for solar, third for nuclear, and sixth for biofuel

South Korea
World's fifth-largest producer of nuclear energy and tenth for solar

Japan
A significant producer of solar (third in world), geothermal, and hydroelectric (both tenth) energy

India
Fourth in the world for wind, sixth for solar, and seventh for hydroelectricity

Philippines
Third-largest geothermal nation

Kenya
Eighth in the world for geothermal; more than 50 percent of Kenya's energy comes from this source

Thailand
Ranked eighth in the world for biofuel

Indonesia
Second in the top 10 of geothermal countries

Australia
Eighth-largest producer of solar energy

New Zealand
Number five in the world for geothermal

Renewable energy

Once fossil fuels have been burned they cannot be replaced. Energy from natural sources that are always replenished—such as wind, sunlight, water, geothermal, and biomass—is known as renewable energy. Most countries are making an effort to increase their usage of this; here are the top 10 producers of renewable energy in 2019.

TOP RENEWABLE ENERGY PRODUCERS

China 758.6
US 264.5
Brazil 141.9
India 128.2
Germany 125.4
Canada 101
Japan 97.5
Italy 55.3
Russia 55.2
France 52.9

Total energy from renewable sources (gigawatt-hours per year)

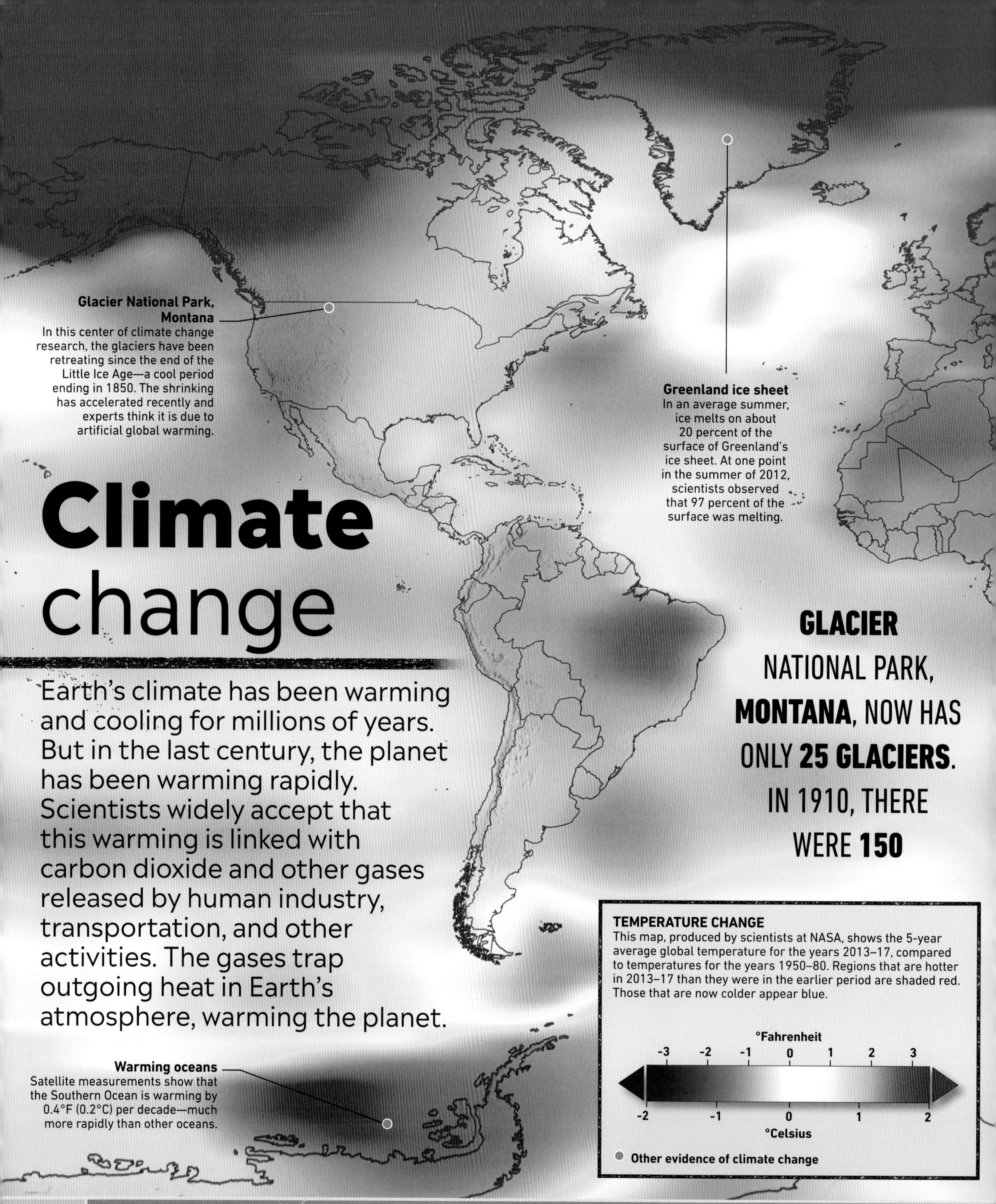

Climate change

Earth's climate has been warming and cooling for millions of years. But in the last century, the planet has been warming rapidly. Scientists widely accept that this warming is linked with carbon dioxide and other gases released by human industry, transportation, and other activities. The gases trap outgoing heat in Earth's atmosphere, warming the planet.

Glacier National Park, Montana
In this center of climate change research, the glaciers have been retreating since the end of the Little Ice Age—a cool period ending in 1850. The shrinking has accelerated recently and experts think it is due to artificial global warming.

Greenland ice sheet
In an average summer, ice melts on about 20 percent of the surface of Greenland's ice sheet. At one point in the summer of 2012, scientists observed that 97 percent of the surface was melting.

Warming oceans
Satellite measurements show that the Southern Ocean is warming by 0.4°F (0.2°C) per decade—much more rapidly than other oceans.

GLACIER NATIONAL PARK, **MONTANA**, NOW HAS ONLY **25 GLACIERS**. IN 1910, THERE WERE **150**

TEMPERATURE CHANGE
This map, produced by scientists at NASA, shows the 5-year average global temperature for the years 2013–17, compared to temperatures for the years 1950–80. Regions that are hotter in 2013–17 than they were in the earlier period are shaded red. Those that are now colder appear blue.

°Fahrenheit
-3 -2 -1 0 1 2 3
-2 -1 0 1 2
°Celsius

Other evidence of climate change

ALL SEVEN OF THE YEARS FROM 2014 TO 2020 RANK AS THE

Arctic sea ice

The entire Arctic region has warmed more during the past decades than any other part of the world, and the most obvious effect is on the sea ice. Sea ice is a layer of frozen sea water, typically 9 ft (3 m) thick, which forms on the surface of the polar oceans. In 2020, Arctic sea ice fell to its second-lowest extent since satellites starting monitoring it in 1979.

Rising sea level

Global sea levels are rising, but satellite data tells us that this is happening more in some places than others. Sea levels around the Philippines, for example, are rising at almost three times the global average, causing the low-lying country to rank on the UN's list of countries most vulnerable to climate change.

Lake shrinkage

Since 1960, Lake Chad has shrunk by 90 percent because the pattern of monsoon rainfall has shifted, giving it less rain. With less rain for their crops, people have also taken much more of the lake's water, shrinking it further. Scientists think the rainfall shift could have been triggered by warming of the surface of the oceans.

Retreating Himalayan glaciers

The shrinking of glaciers in the Himalayas north of India has been blamed on the Asian Brown Cloud—a haze of sooty particles released by south Asian cities. Even though the cloud blocks some sunlight reaching Earth, cooling the land below, it has a warming effect overall, because it absorbs and traps heat energy like carbon dioxide does.

Melting glacier

Muir Glacier in Alaska has been shrinking for more than 80 years. It has retreated by more than 7 miles (12 km) and is 2,600 ft (800 m) thinner—this shrinkage is shown in the photos below. The glacier is now out of sight from this angle.

August 13, 1941

August 31, 2004

Global sea level

Since 1993, the global sea level has steadily increased by about 0.1 in (3 mm) per year, as measured by satellite. During the 100-year period before 1993, sea levels rose by an average of only 0.07 in (1.7 mm) every year.

Sea level rise since 1993—

Change in sea level (in)

Change in sea level (mm)

1994 2000 2005 2010 2015 2020

Alert, Canada

Queen Maud Gulf Migratory Bird Sanctuary, Arctic Canada

Kluane/Wrangell-St. Elias/ Glacier Bay/Tatshenshini-Alsek, Alaska and British Columbia
Home to some of the world's most spectacular glaciers.

Northeast Greenland National Park
Once the world's largest protected area, mostly made up of the Greenland Ice Sheet.

Charlie-Gibbs Marine Protected Area, Atlantic Ocean

Yellowstone National Park, Wyoming
The first national park in the world, founded in 1872. The large alpine meadows and grass prairies provide the ideal habitat for the large herds of bison living in the park.

Northern Canada
Permafrost (permanently frozen soil) makes this vast region inaccessible to people, preserving the Arctic tundra plains for the wolves and caribou.

Papahānaumokuākea Marine National Monument
Hawaii, 585,242 sq mi (1,508,000 sq km).

Sahara desert
World's largest hot desert. Supports little human life other than in scattered oases.

Pacific Remote Islands Marine National Monument
Central Pacific Ocean, 490,543 sq mi (1,270,500 sq km).

Galápagos Islands

Aïr and Ténéré Natural Reserves, Niger

Marae Moana
Cook Islands, 762,938 sq mi (1,976,000 sq km).

Amazon rainforest
The north and west of this great forest have few or no roads and are far from human impact. Some areas are flooded to great depth every year. Some parts have never been logged and are "pristine."

Jaú National Park, Amazonas, Brazil
One of the largest protected rainforest areas in the world and the largest in the Amazon basin. The park includes the entire Jaú River, where the water is black from minerals in dissolved organic matter.

WILDERNESS AREAS
The map shows the level of human influence across the world. The colors are based on the "wilderness value," which measures how far any one place is from permanent human settlements, roads, and man-made structures. This measure of remoteness from human development shows how much wilderness is left.

Key

High wilderness   Low wilderness

PROTECTED AREAS OF THE WORLD
The blue areas on the map show some of the world's protected areas of wilderness. Damaging activities, such as hunting and mining, are usually banned. The areas include wildlife reserves, national parks, marine parks, and more.

⬤ Top 5 largest protected areas

ABOUT **50 PERCENT** OF THE WORLD'S **PEOPLE** LIVE ON JUST **1 PERCENT** OF THE **LAND**

Yugyd Va National Park, Russia
One of the largest national parks in Europe, made up of forests in the Northern Ural Mountains.

Siberia
The northeastern parts of Russia are cold, remote, and largely untouched forests, mountains, and tundra.

Pelagos Sanctuary for Mediterranean Marine Mammals

Great Siberian Polynia, Russia

Lake Baikal World Heritage Site, Russia

Qiangtang Nature Reserve, China
The Tibetan Plateau is remote and sparsely populated. From the most remote point, it is a three-week trip to the nearest cities of Lhasa or Korla—one day by car and the remaining 20 on foot.

Outback, Australia
"Outback" describes the hot, dry parts of Australia where very few people live, many of whom are Aboriginals. Several highways pass through even the most remote areas.

Selous Game Reserve Tanzania

Kavango-Zambezi Transfrontier Conservation Area
Home to a varied wildlife population. Victoria Falls, between Zambia and Zimbabwe, lies at the center.

Chagos Marine Protected Area, Indian Ocean
A British-controlled cluster of 55 tiny islands, surrounded by a vast marine reserve.

Natural Park of the Coral Sea
New Caledonia, 499,230 sq mi (1,293,000 sq km).

Great Barrier Reef Marine Park, Australia
World's largest coral reef system.

Ross Sea Region Marine Protected Area
Antarctica, 598,458 sq mi (1,550,000 sq km).

Wilderness

Wildernesses are the last places that have been largely unchanged by humans. Indigenous peoples sometimes live in these undeveloped areas, where their lifestyles sometimes impact little on the landscape and wildlife.

THE EARTH'S MOST NORTHERLY PERMANENTLY INHABITED PLACE.

Engineering and technology

Reaching for the sky
The Burj Khalifa, the world's tallest building, can be seen in the distance in this view of fog-bound Dubai, the largest city in the United Arab Emirates.

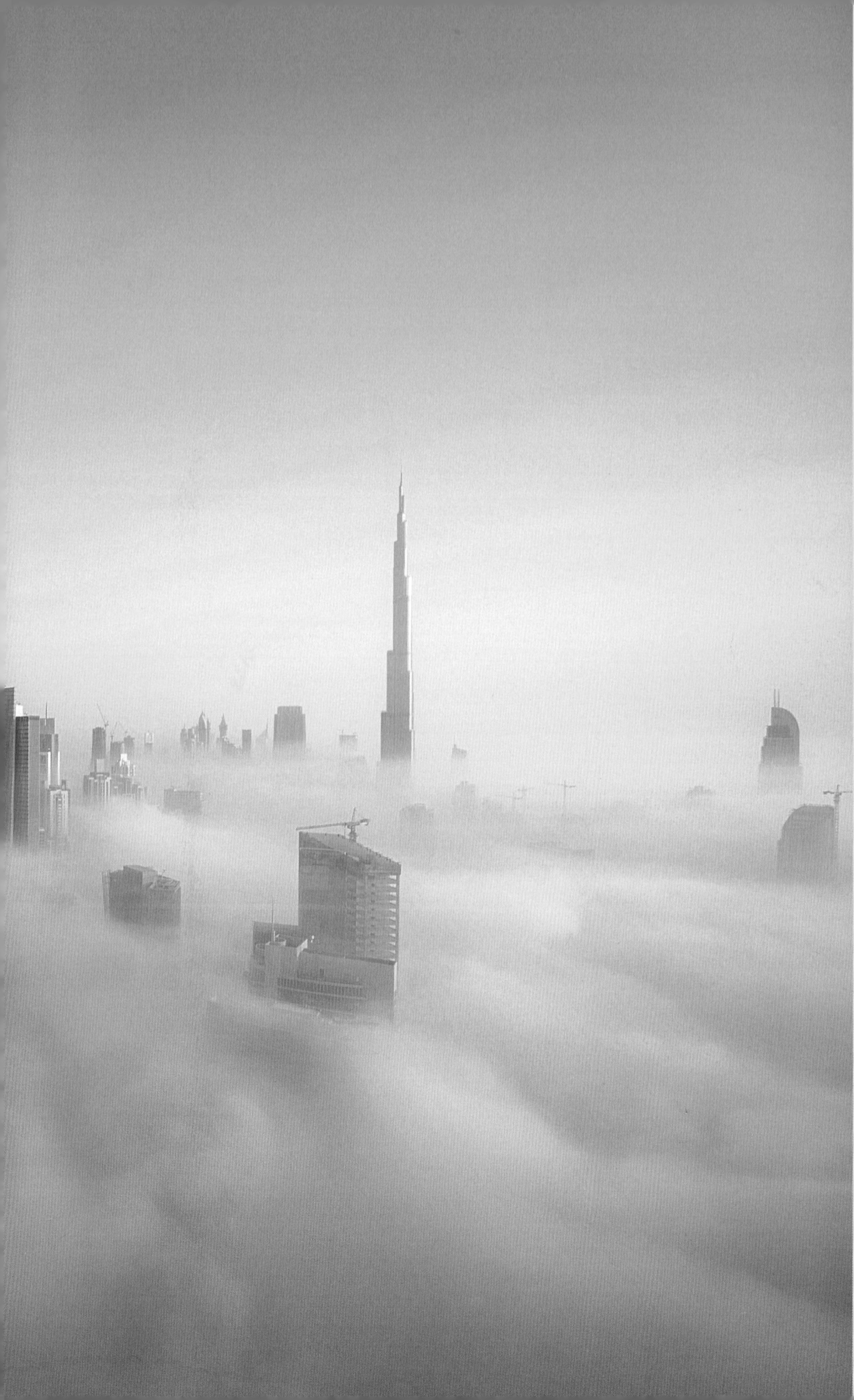

Introduction

Engineering and technology enable humans to achieve amazing feats. We build skyscrapers that reach toward the clouds, bridges that span great canyons, and tunnels that pierce mountains and travel under the sea. Our computer networks and transportation systems keep people and places connected. We can even explore other planets.

World in motion

Transportation has shrunk our world. Thanks to jet airliners, superhighways, and high-speed rail routes, we can go on long-distance journeys that would have been unthinkable just a few decades ago. This transportation revolution began with the invention of the railroad at the start of the 19th century, and it has continued at speed ever since.

Train collects electricity from power cables suspended above the track.

High-speed electric locomotive
Launched in 1999, the Velaro is now in service in Germany, Spain, France, the UK, China, Russia, Belgium, Turkey, and the Netherlands. It is powered by electricity and can reach speeds of more than 218 mph (350 kph).

Shrinking technology

Few, if any, areas of technology have advanced faster than computing. ENIAC, developed by the US Army in 1946, was the first general-purpose programmable electronic computer. ENIAC contained more than 100,000 components. Since then, electronic components have become smaller and smaller. A modern laptop computer is controlled by a tiny microchip that may be etched with more than a billion components.

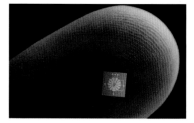

Modern marvel
This tiny computer, just 0.04 in (1 mm) square, is implanted into the eye to help people with the disease glaucoma.

Enormous ancestor
ENIAC weighed 33 tons and occupied an entire room. Operators programmed ENIAC by plugging and unplugging cables and adjusting switches.

FUGAKU WAS CROWNED THE WORLD'S FASTEST COMPUTER IN 2020, AND

Early steam engine
Puffing Billy is the world's oldest surviving steam locomotive. Built in 1813 to haul coal in northern England, it had a top speed of about 6 mph (10 kph).

Coal carried in the tender was burned to heat water in the boiler and produce steam to drive the wheels.

Bullet-shaped nose enables locomotive to cut through the air more easily, increasing speed.

Construction

A steel-and-concrete building revolution began in the late 19th century. Frames made of steel girders allowed taller structures to be built, and the invention of reinforced concrete—concrete with steel rods set into it—introduced an amazingly strong, durable new material. Together, steel and reinforced concrete gave birth to the modern skyscraper, changing the face of the world's cities.

- **Ancient concrete**
  The Romans were experts in building with concrete. It was used in the construction of the Colosseum and the Pantheon in Rome.

- **World's oldest skyscraper city?**
  Shibam, in Yemen, has about 500 high-rise apartment buildings made of mud brick, most dating from the 16th century.

- **First steel-framed skyscraper**
  Completed in 1885, the innovative 10-story Home Insurance Building in Chicago, Illinois, used a steel frame to support the walls.

- **Reinforced first**
  The first skyscraper built with reinforced concrete was the 15-story Ingalls Building, in Cincinnati, Ohio, erected in 1903.

Manhattan, then and now
The Brooklyn Bridge spans New York's East River. The view across to Manhattan Island has changed dramatically since the bridge opened in 1883, and it now bristles with skyscrapers.

Infrastructure

The built and engineered systems that we rely on every day—from sewers and telecommunication networks to power lines, railroads, and roads—are collectively known as infrastructure. Without such systems, our modern way of life would be impossible.

- **First telephone exchange**
  The first commercial exchange to connect callers was built in New Haven, Connecticut, in 1878.

- **Intercity railroad**
  Opened in 1830, the Manchester to Liverpool route in England was the first intercity railroad.

Ulm–Stuttgart autobahn, 1950
Germany was a pioneer of the freeway, or autobahn, in the 1930s. Cars did not clog the roads until much later!

IN **2019**,
HARTSFIELD-JACKSON,
ATLANTA, AVERAGED
2,569 FLIGHTS
PER DAY

Top 10 busiest passenger airports 2019

About 9.1 billion air passengers passed through the world's top 100 airports in 2019. The world's busiest airport, Hartsfield-Jackson International in Atlanta, Georgia, averaged 275,000 passengers per day in 2019, and handled more than 904,301 flights during the year. The industry declined drastically in 2020, however, when COVID-19 brought the world to a standstill.

RANK	AIRPORT	PASSENGERS PER YEAR
1	Hartsfield–Jackson Atlanta International, US	110,531,300
2	Beijing Capital International, China	100,011,438
3	Los Angeles International, US	88,068,013
4	Dubai International, Dubai	86,396,757
5	Tokyo International, Japan	85,505,054
6	O'Hare International, Chicago, US	84,649,115
7	London Heathrow, United Kingdom	80,888,305
8	Shanghai Pudong International, China	76,153,455
9	Paris Charles de Gaulle, France	76,150,009
10	Dallas Fort Worth International, US	75,066,956

Air traffic

Air-traffic controllers have a tough job ensuring safe routes, takeoffs, and landings for the thousands of planes that crisscross our skies each day. This map shows nearly 6,000 routes carrying scheduled commercial traffic.

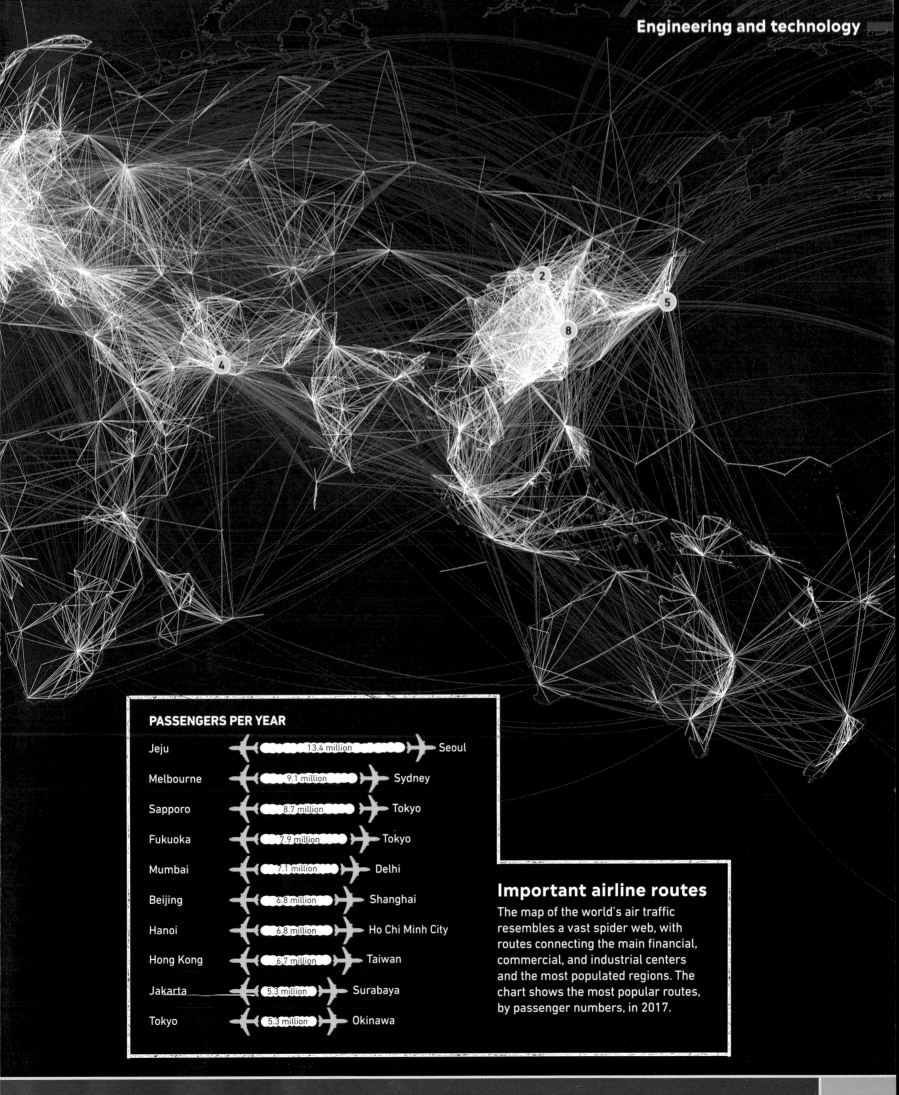

PASSENGERS PER YEAR

Jeju	13.4 million	Seoul
Melbourne	9.1 million	Sydney
Sapporo	8.7 million	Tokyo
Fukuoka	7.9 million	Tokyo
Mumbai	7.1 million	Delhi
Beijing	6.8 million	Shanghai
Hanoi	6.8 million	Ho Chi Minh City
Hong Kong	6.7 million	Taiwan
Jakarta	5.3 million	Surabaya
Tokyo	5.3 million	Okinawa

Important airline routes

The map of the world's air traffic resembles a vast spider web, with routes connecting the main financial, commercial, and industrial centers and the most populated regions. The chart shows the most popular routes, by passenger numbers, in 2017.

IN 2019. IT IS PREDICTED THAT THERE WILL BE 51.71 MILLION IN 2030.

117

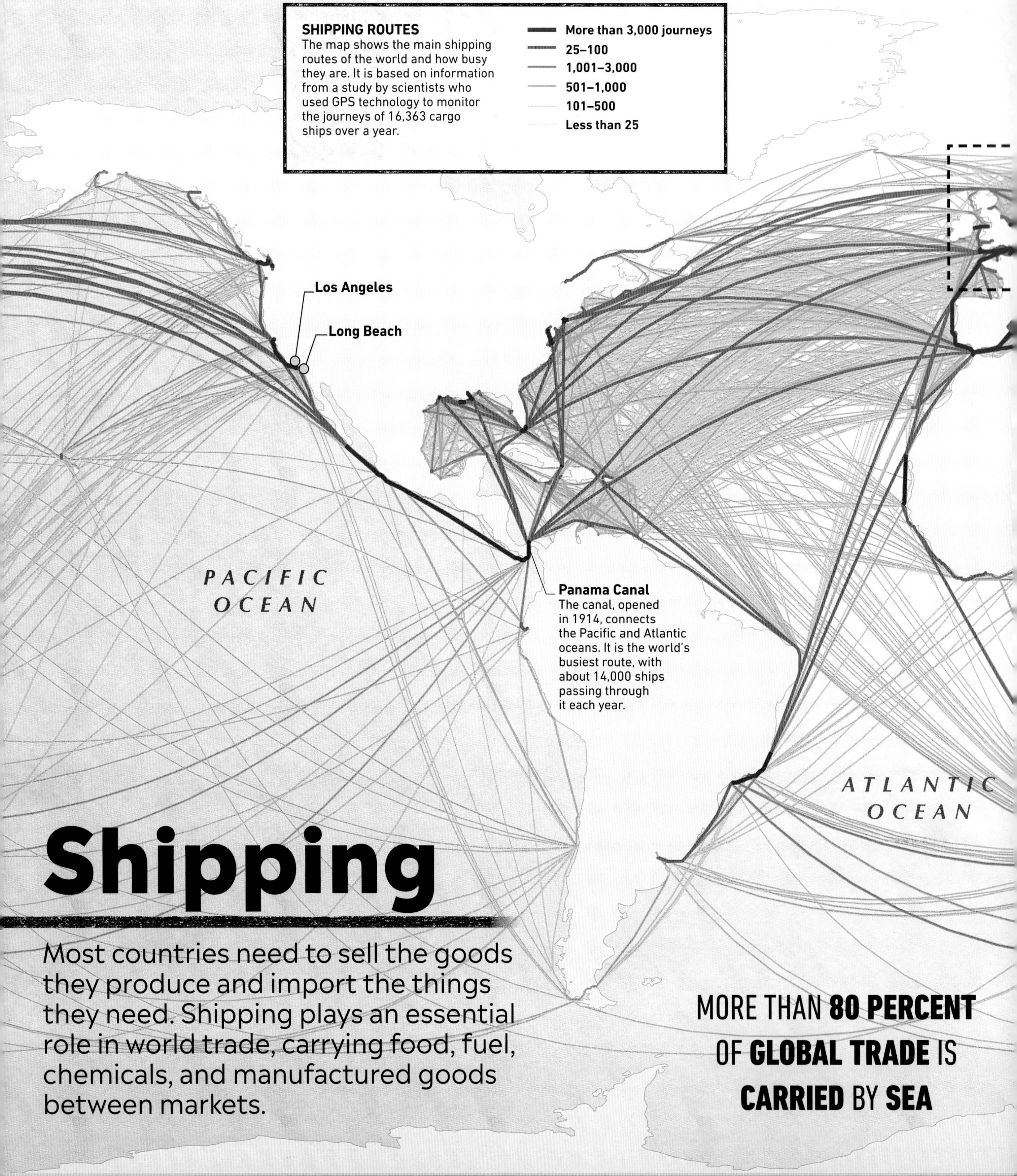

The map shows the main shipping routes of the world and how busy they are. It is based on information from a study by scientists who used GPS technology to monitor the journeys of 16,363 cargo ships over a year.

▬▬▬	More than 3,000 journeys
▬▬▬	25–100
▬▬▬	1,001–3,000
▬▬▬	501–1,000
┈┈┈	101–500
─────	Less than 25

Los Angeles

Long Beach

PACIFIC OCEAN

Panama Canal
The canal, opened in 1914, connects the Pacific and Atlantic oceans. It is the world's busiest route, with about 14,000 ships passing through it each year.

ATLANTIC OCEAN

Shipping

Most countries need to sell the goods they produce and import the things they need. Shipping plays an essential role in world trade, carrying food, fuel, chemicals, and manufactured goods between markets.

MORE THAN **80 PERCENT** OF **GLOBAL TRADE** IS **CARRIED** BY **SEA**

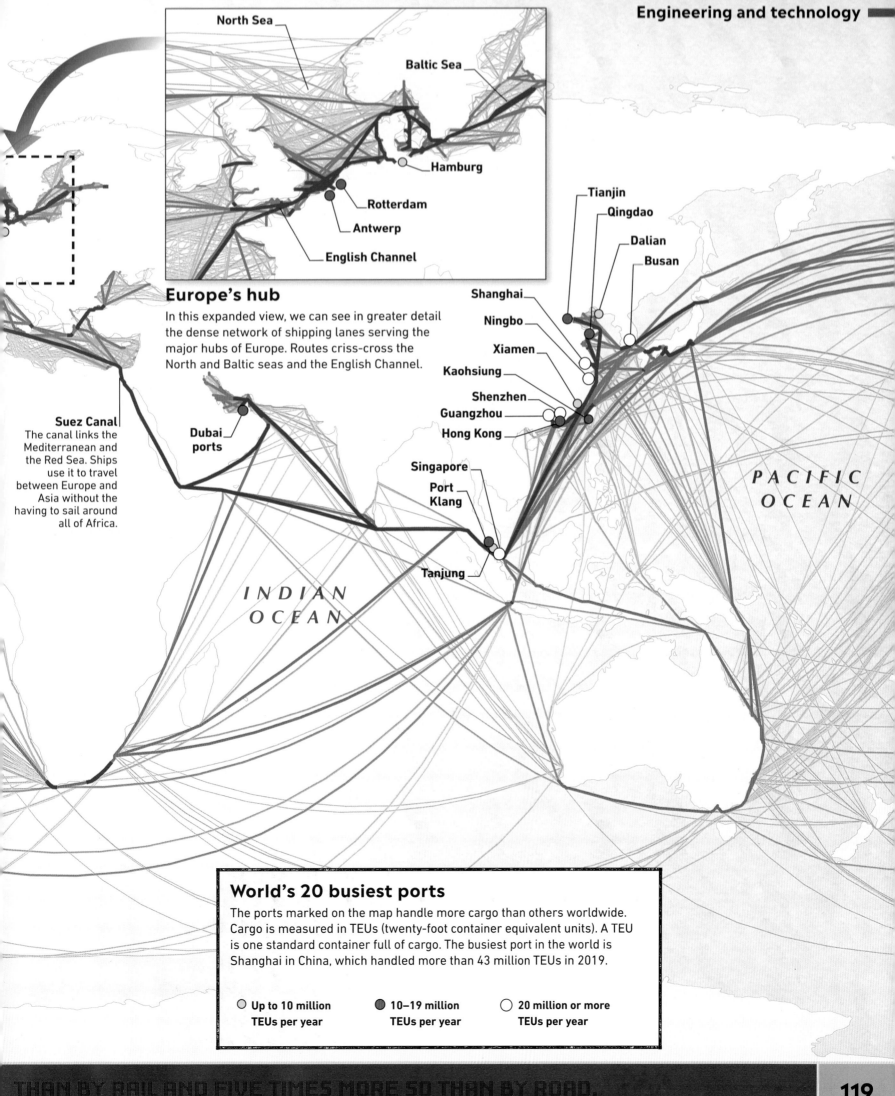

North Sea

Baltic Sea

Hamburg

Rotterdam

Antwerp

English Channel

Europe's hub

In this expanded view, we can see in greater detail the dense network of shipping lanes serving the major hubs of Europe. Routes criss-cross the North and Baltic seas and the English Channel.

Tianjin

Qingdao

Dalian

Busan

Shanghai

Ningbo

Xiamen

Kaohsiung

Shenzhen

Guangzhou

Hong Kong

Suez Canal
The canal links the Mediterranean and the Red Sea. Ships use it to travel between Europe and Asia without the having to sail around all of Africa.

Dubai ports

Singapore

Port Klang

Tanjung

PACIFIC OCEAN

INDIAN OCEAN

World's 20 busiest ports

The ports marked on the map handle more cargo than others worldwide. Cargo is measured in TEUs (twenty-foot container equivalent units). A TEU is one standard container full of cargo. The busiest port in the world is Shanghai in China, which handled more than 43 million TEUs in 2019.

○ Up to 10 million TEUs per year

● 10–19 million TEUs per year

○ 20 million or more TEUs per year

Railroads

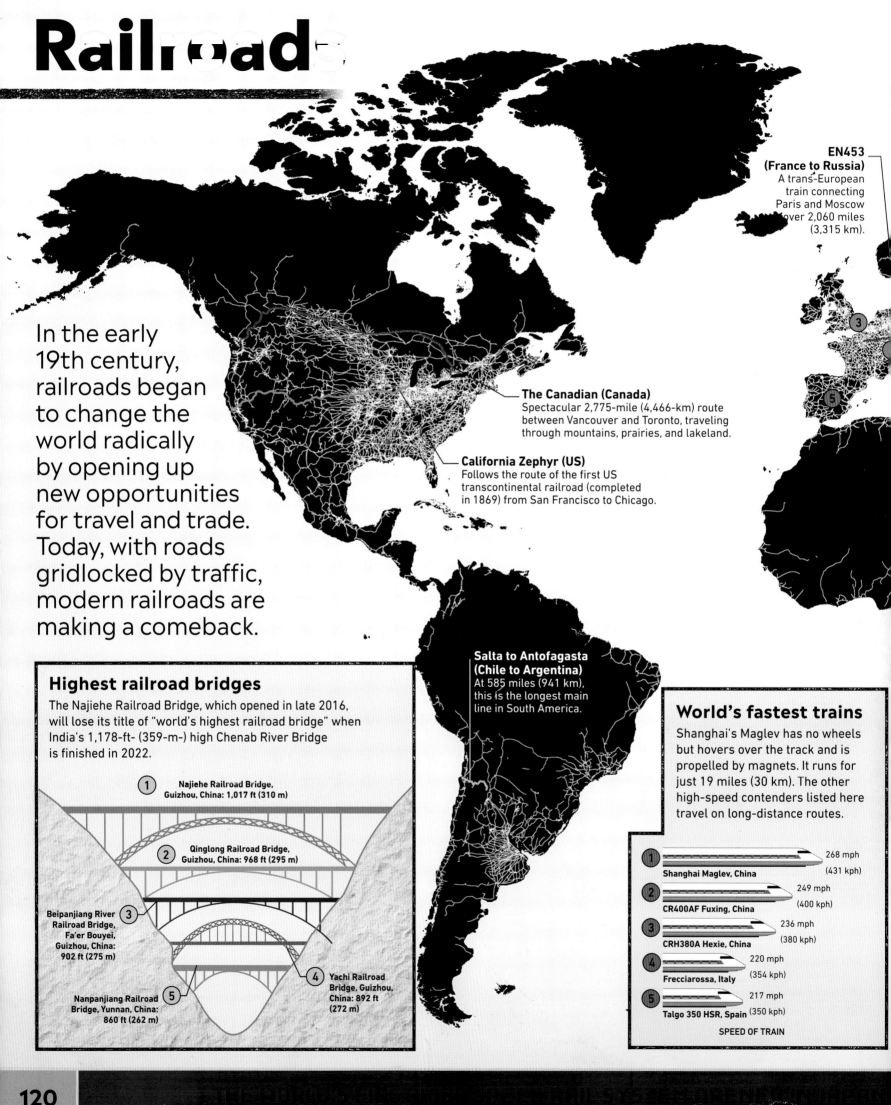

In the early 19th century, railroads began to change the world radically by opening up new opportunities for travel and trade. Today, with roads gridlocked by traffic, modern railroads are making a comeback.

EN453 (France to Russia)
A trans-European train connecting Paris and Moscow over 2,060 miles (3,315 km).

The Canadian (Canada)
Spectacular 2,775-mile (4,466-km) route between Vancouver and Toronto, traveling through mountains, prairies, and lakeland.

California Zephyr (US)
Follows the route of the first US transcontinental railroad (completed in 1869) from San Francisco to Chicago.

Salta to Antofagasta (Chile to Argentina)
At 585 miles (941 km), this is the longest main line in South America.

Highest railroad bridges

The Najiehe Railroad Bridge, which opened in late 2016, will lose its title of "world's highest railroad bridge" when India's 1,178-ft- (359-m-) high Chenab River Bridge is finished in 2022.

1. Najiehe Railroad Bridge, Guizhou, China: 1,017 ft (310 m)
2. Qinglong Railroad Bridge, Guizhou, China: 968 ft (295 m)
3. Beipanjiang River Railroad Bridge, Fa'er Bouyei, Guizhou, China: 902 ft (275 m)
4. Yachi Railroad Bridge, Guizhou, China: 892 ft (272 m)
5. Nanpanjiang Railroad Bridge, Yunnan, China: 860 ft (262 m)

World's fastest trains

Shanghai's Maglev has no wheels but hovers over the track and is propelled by magnets. It runs for just 19 miles (30 km). The other high-speed contenders listed here travel on long-distance routes.

1. Shanghai Maglev, China — 268 mph (431 kph)
2. CR400AF Fuxing, China — 249 mph (400 kph)
3. CRH380A Hexie, China — 236 mph (380 kph)
4. Frecciarossa, Italy — 220 mph (354 kph)
5. Talgo 350 HSR, Spain — 217 mph (350 kph)

SPEED OF TRAIN

Trans-Siberian Railroad (Russia)
The world's longest rail journey passes through seven time zones as it runs 5,771 miles (9,288 km) from Moscow in the west to Vladivostock on the Pacific coast.

THE WORLD HAS 620,000 MILES (1 MILLION KM) OF **RAIL TRACK**

Guangzhou to Lhasa (China)
The Tanggula Pass section is the world's highest track, at 16,640 ft (5,072 m).

Alexandria to Aswan (Egypt)
Traveling via Cairo and Luxor, this line follows the Nile Valley, with its ancient pyramids and temples.

Dibrugarh to Kanyakumari (India)
Longest route in India, at 4,286 km (2,657 miles).

Indian Pacific (Australia)
This 2,704-mile (4,352-km) route, links Sydney on the east coast with Perth in the west.

Five longest railroad tunnels

In 2016, the Gotthard Base Tunnel—a 35-mile- (57-km-) long tunnel that travels beneath the Swiss Alps—surpassed the Seikan Tunnel to become the world's longest railroad tunnel.

Blue Train, South Africa
A luxurious train that runs from Cape Town to Pretoria through vineyards, mountains, and the arid landscape of the Karoo.

Yulhyeon Tunnel, Gyeonggi South Korea:
31.3 miles (50.3 km)

Songshan Lake Tunnel, Dongguan, China:
24.1 miles (38.8 km)

Channel Tunnel, English Channel:
31.5 miles (50.4 km)

Gotthard Base Tunnel, Swiss Alps, Switzerland:
35.5 miles (57.1 km)

Seikan Tunnel, Tsugaru Strait, Japan:
33.5 miles (53.8 km)

Dempster Highway Extension
An ice road built on the frozen Mackenzie River and Arctic Ocean, it provides a winter route to the isolated community of Tuktoyaktuk.

Mountain roads and passes

① Trollstigen, Norway
This dramatic road's name means "Trolls' ladder." It has 11 hairpin bends, which wind up the steep mountainside.

② Stelvio Pass, Italy
One of the highest roads in the Alps, its 60 hairpin bends provide a challenge for both drivers and bicyclers.

③ Khardung La, India
This famously high mountain pass in the Ladakh part of Kashmir was built in 1976 and opened to motor vehicles in 1988.

④ Semo La, Tibet, China
Possibly the highest vehicle-accessible pass in the world, it was reliably measured in 1999 at 18,258 ft (5,565 m).

⑤ Irohazaka Winding Road
Each of the 48 hairpin turns on this route in Japan is labeled with one of the 48 characters of the Japanese alphabet.

Tibbit to Contwoyto Winter Road
An ice road built over frozen lakes, it is open for about 10 weeks from late January each year.

Pacific Coast Highway
This world-famous route hugs the California coast from Orange County in the south to the forests of giant redwood trees in the north.

Bonn-Köln Autobahn
Built in 1932, it was the first road designed exclusively for cars, with divided lanes and no intersections with other roads.

Cabot Trail
Looping around the northern tip of Cape Breton Island, Nova Scotia, and named after 16th-century Italian explorer, John Cabot.

Route 66
A 2,448-mile (3,940-km) road that follows the historic route taken by migrants to California during the Great Depression.

Natchez Trace Parkway
A route used by Native Americans and their animals for thousands of years before the modern road was built.

Darién Gap, Panama
A stretch of rainforest that breaks the Pan-American Highway's route.

Pan-American Highway
About 29,800 miles (48,000 km) long, it runs through 18 countries, from Alaska to the southern tip of Argentina.

Yungas Road, Bolivia
A single-track mountain road heavily used by trucks but with unprotected sheer drops of 1,970 ft (600 m). Up to 300 travelers are killed on the route every year.

World's busiest roads

① Ontario Highway 401, Canada
The busiest highway in North America—more than 440,000 vehicles pass through the Toronto section every day. It is also one of the widest in the world—some sections of the route have 18 lanes.

② Interstate 405, California
Runs north from the city of Irvine in Orange County to San Fernando, a route that is known as the northern segment of the San Diego Freeway. This freeway is the busiest and most congested in the US, carrying up to 379,000 vehicles a day.

HIGHWAY 401, ONTARIO, CANADA

Roads

The planet is now more accessible by road than it has ever been. There are about 65 million miles (104 million km) of roads on Earth, from multilane urban freeways to seasonal ice roads made from frozen lakes and seas.

Estonian Islands
Ice roads between islands and the mainland are only opened to traffic when the ice is 8.7 in (22 cm) thick along the entire route.

Siberia
Siberia has few permanent roads, partly because it is so difficult to build stable foundations on the permafrost soil.

Road of Bones
The M56 Kolyma Highway passes through the coldest inhabited places on Earth, with winter temperatures dropping below −58° F (−50° C).

Karakoram Highway
One of the world's highest roads at an altitude of 15,397 ft (4,693 m), it connects China and Pakistan.

THE **GEORGE WASHINGTON BRIDGE** IN NEW YORK CARRIES 104 MILLION **VEHICLES** EVERY YEAR

Milford Road
Meanders through the stunning scenery of New Zealand's Milford Sound.

Garden Route
Runs along the South African coast from Cape Town to Port Elizabeth.

Great Ocean Road
Following a beautiful seaside route, this road is a memorial to the Australians who died in World War I.

Record road bridges

① Millau Viaduct
This French bridge is the tallest in the world. One mast is 1,125 ft (343 m) tall—taller than the Eiffel Tower.

② Beipanjiang Bridge
With the road 1,850 ft (565 m) above the Beipan River Canyon in China, this is the world's highest bridge.

③ Bang Na Expressway
This 34-mile- (55-km-) long six-lane elevated highway in Thailand is the world's longest road bridge.

④ Jiaozhou Bay Bridge
The world's longest road bridge crossing water, it is supported by 5,238 massive concrete pillars.

⑤ Lake Pontchartrain Causeway
Two parallel bridges 24 miles (38 km) long, near New Orleans, Louisiana.

⑥ Akashi-Kaikyō Bridge
The world's longest suspension bridge, it has 190,000 mi (300,000 km) of steel cables and connects two Japanese islands.

MILLAU VIADUCT, FRANCE

KEY
Roads can be paved (covered stones, brick, concrete, tarmac, or another hard surface), or unpaved. Paving makes a road more durable and weatherproof.

—— **Famous roads**
—— **Scenic routes**
—— **Ice roads**

SOUTH SUDAN IS THE SAME SIZE, BUT HAS JUST 186 MILES (300 KM).

123

CN Tower
1,815 ft (553 m)
Toronto, Canada
1976

One World Trade Center
1,776 ft (541 m)
New York, New York
2013

Willis Tower
1,450 ft (442 m)
Chicago, Illinois
1973

Commerzbank
850 ft (259 m)
Frankfurt, Germany
1997

Transamerica Pyramid
850 ft (260 m)
San Francisco, California
1972

The Shard
1,016 ft (310 m)
London, UK
2012

Central Park Tower
1,550 ft (472 m)
New York City, New York
2021

Torre Caja Madrid
820 ft (250 m)
Madrid, Spain
2008

Trump Ocean Club
961 ft (293 m)
Panama City, Panama
2011

Empire State Building
1,250 ft (381 m)
New York City, New York
1931

John Hancock Center
1,129 ft (344 m)
Chicago, Illinois
1969

The Eiffel Tower
1,063 ft (324 m)
Paris, France
1889

Gran Torre Santiago
300 m (984 ft)
Santiago, Chile
2012

Great Pyramid of Giza
481 ft (147 m)
El Giza, Egypt
c.2560–2540 BCE

Carlton Center
732 ft (223 m)
Johannesburg, South Africa
1973

Makka Royal Clock Tower Hotel
1,972 ft (601 m)
Makkah, Saudi Arabia
2012

Tallest buildings

From ancient pyramids to today's high-tech hotels, powerful people have shown off their status through impressive buildings. As technology improves, the towers get taller.

Big buildings

To be called a building, a tower must be inhabitable (offices or homes). Buildings do not include "supported structures" such as guyed (tethered) masts. Buildings may be measured to their architectural top, as on these pages, or to the tip of any masts or aerials. Here are some of the world's tallest.

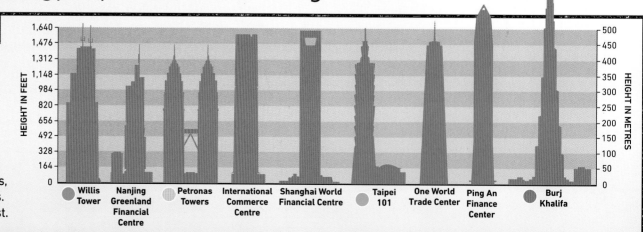

2,717 ft (828 m)

HEIGHT IN FEET: 1,640 · 1,476 · 1,312 · 1,148 · 984 · 820 · 656 · 492 · 328 · 164 · 0

HEIGHT IN METRES: 500 · 450 · 400 · 350 · 300 · 250 · 200 · 150 · 100 · 50 · 0

Willis Tower · Nanjing Greenland Financial Centre · Petronas Towers · International Commerce Centre · Shanghai World Financial Centre · Taipei 101 · One World Trade Center · Ping An Finance Center · Burj Khalifa

THE BURJ KHALIFA HAS **163** FLOORS LINKED BY **57** DOUBLE-DECKER LIFTS

Ostankino Tower
1,770 ft (540 m)
Moscow, Russia
1967

Mercury City Tower
1,112 ft (339 m)
Moscow, Russia
2012

International Commerce Center
1,588 ft (484 m)
Hong Kong
2010

Oriental Pearl Tower
1,535 ft (468 m)
Shanghai, China
1994

Shanghai World Financial Center
1,614 ft (492 m)
Shanghai, China
2008

Shanghai Tower
2,073 ft (632 m)
Shanghai, China
2014

Tianjin CTF Finance Center
1,739 ft (530 m)
Tianjin, China
2018

Ping An Finance Center
1,965 ft (599 m)
Shenzhe, China
2017

Milad Tower
1,427 ft (435 m)
Tehran. Iran
2007

Burj Khalifa
2,717 ft (828 m)
Dubai, UAE
2010

Tokyo Sky Tree
2,080 ft (634 m)
Tokyo, Japan
2011

Taipei 101
1,670 ft (509 m)
Taipei, Taiwan
2004

Busan Lotte Tower
1,674 ft (510.2 m)
Busan, South Korea
2015

Petronas Towers
1,483 ft (452 m)
Kuala Lumpur,
Malaysia
1998

Canton Tower
1,969 ft (600 m)
Guangzhou, China
2010

Q1
1,060 ft (323 m)
Gold Coast,
Australia
2005

Unsupported towers

Unlike buildings, these structures don't contain offices, homes, or stores. They are observation and communications towers.

● **Tokyo Sky Tree**
This communications tower overtook the Canton Tower in 2011 to become the world's tallest.

● **Canton Tower**
Canton is the former name of Guangzhou, where this tower was completed in 2010.

● **CN Tower**
More than 2 million people visit this tower's glass-floored observation deck every year.

● **Ostankino Tower**
This broadcasting tower was the world's first free-standing structure over 1,640 ft (500 m) tall.

● **Oriental Pearl Tower**
There are 11 spheres in the design of this TV tower, which has 15 observation levels.

Record-breaking buildings

The record for the tallest building (a structure that must be inhabitable) is a fiercely contested prize. These five have all won it.

● **Burj Khalifa, 2010–present**
This building has broken all records, including the tallest building and tallest unsupported structure.

● **Taipei 101, 2004–10**
The world's tallest building until the Burj Khalifa was built, Taipei 101 has 101 floors above ground.

● **Petronas Towers, 1998–2004**
These office blocks were the tallest buildings until 2004. They are still the tallest twin towers.

● **Willis Tower, 1973–98**
Formerly known as the Sears Tower, this 108-story skyscraper towers above Chicago.

● **Empire State Building, 1931–72**
This was the first building in the world to have more than 100 stories—it has 102. It was the tallest building for 40 years.

Internet
connections

The Internet has revolutionized the way we live our lives. At the click of a mouse, we can instantly exchange news, ideas, and images with people on the other side of the world, and we can buy or sell goods without having to leave our homes.

The Internet in a minute

Today, there are more than three times as many computers, phones, and other devices connected to the Internet as there are people in the world. As a result, an incredible amount of Internet activity can occur in just one minute.

4.7 million videos viewed

4.1 million Google searches

59 million messages sent

347,222 stories viewed

764,000 hours of Netflix watched

$1.1 million in online sales

BY **OCTOBER 2012,**
THERE WERE AT LEAST
10 BILLION WEB PAGES

Internet connection speed

Nowadays, most Internet connections are broadband, provided
by digital phone lines, satellites, or fiber-optic cables. These are
much faster than the connections that used to be common,
provided via ordinary phone lines and a modem. Following
the huge rise of working from home due to the COVID-19
pandemic, Internet speed has never been more
important. Here is a selection of the download speeds
in different countries in 2020. Internet users in
Liechtenstein had the world's fastest broadband,
with an average peak download speed of
just under 230 megabits per second.

200th: CHINA 2.09
101st: INDIA 13.46
76th: GREENLAND 18.65
47th: UNITED KINGDOM 37.82
20th: UNITED STATES 71.30
5th: LUXEMBOURG 118.05
4th: GIBRALTAR 183.1
3rd: ANDORRA 213.41
2nd: JERSEY 218.37
1st: LIECHTENSTEIN 229.98

PEAK CONNECTION SPEED (MEGABITS PER SECOND) AND WORLD RANKING

A web of connectivity

The map shows how the world's
cities are connected by the Internet—
the brighter the area, the more
connections there are. Connections
are not the same as users. Many
people, for example, use a single
connection in an Internet café.

—— Lines represent Internet
connections between cities

Satellites and space junk

The first satellite, *Sputnik 1*, was launched by the Soviet Union in 1957. Since then, thousands of satellites and millions of other objects have accumulated around Earth, creating a serious hazard for space travel.

Geosynchronous ring
This ring-shaped concentration of satellites appears more than 22,200 miles (35,700 km) above Earth's equator. It exists because it is extremely useful for a satellite to "hover" above a point on Earth's turning surface.

High-speed danger

The pattern of spots shows the strikes collected during the entire NASA Space Shuttle program, from 1983–2002. The vast majority of space debris is less than 0.5 in (1 cm) across and includes specks of solid rocket fuel and flakes of paint. But even dust acts like tiny bullets at speeds of up to 26,000 mph (42,000 kph).

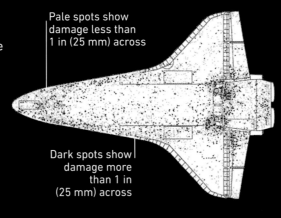

Pale spots show damage less than 1 in (25 mm) across

Dark spots show damage more than 1 in (25 mm) across

AT LEAST **10 MILLION** PIECES OF ARTIFICIAL DEBRIS ARE NOW IN **EARTH ORBIT**

KEY
The image shows 22,300 objects monitored by the ESA Space Debris team by radar and telescopes.

Satellites—mostly dead. About 2,300 operational

Spent rockets

Mission waste (nuts, gloves, lost items)

Debris from explosions and collisions

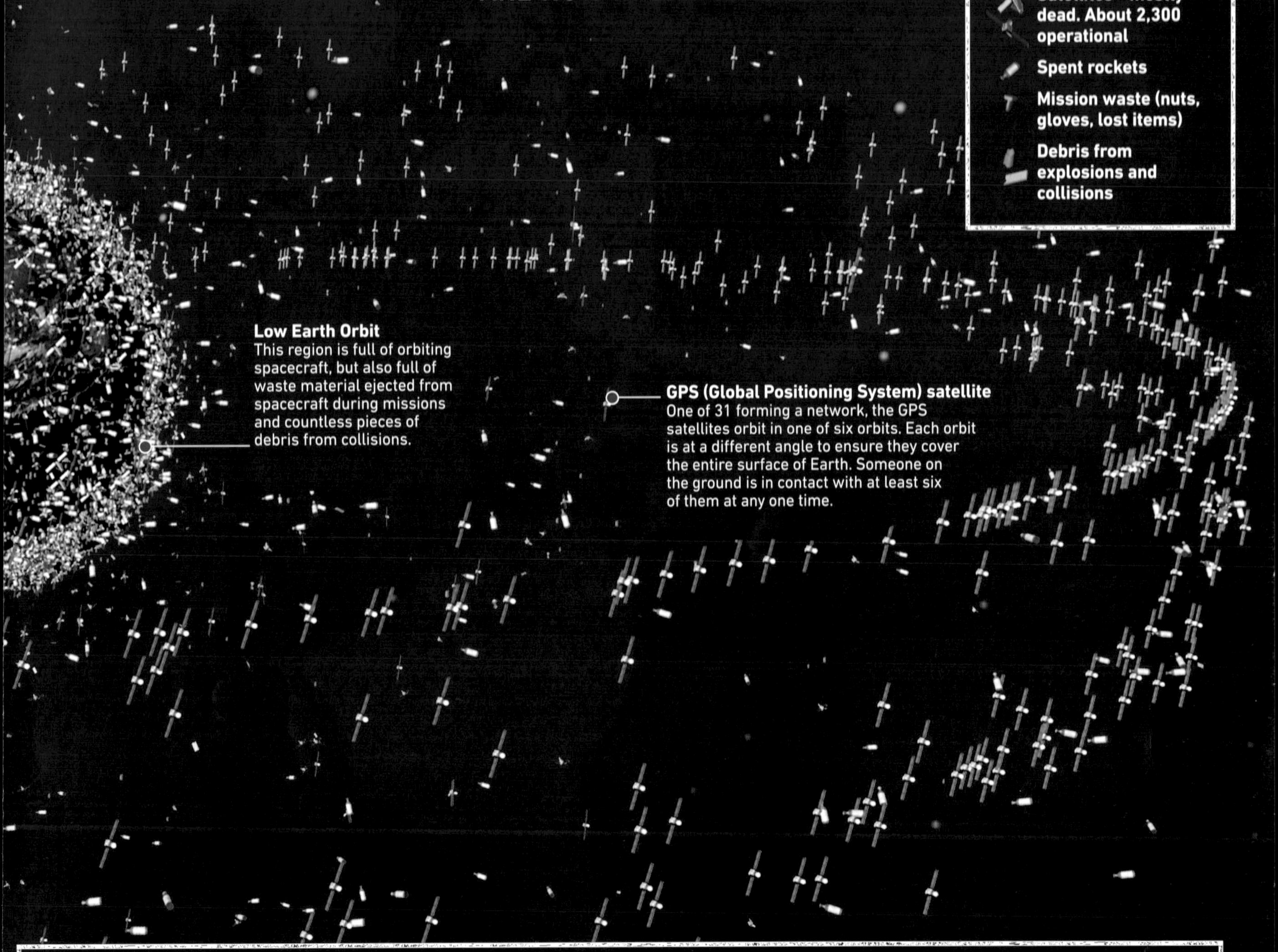

Low Earth Orbit
This region is full of orbiting spacecraft, but also full of waste material ejected from spacecraft during missions and countless pieces of debris from collisions.

GPS (Global Positioning System) satellite
One of 31 forming a network, the GPS satellites orbit in one of six orbits. Each orbit is at a different angle to ensure they cover the entire surface of Earth. Someone on the ground is in contact with at least six of them at any one time.

How high are satellites?

Most objects launched into space are in Low Earth Orbit (LEO). At the lowest LEOs (99 miles / 160 km) objects circle Earth in 87 minutes at 17,470 mph (28,100 kph). Certain orbits are particularly useful. Image-taking satellites use polar sun-synchronous orbits, which pass the equator at the same local time on every pass, so the shadows are the same.

Geosynchronous orbit
22,236 miles (35,786 km)
Satellites at this height orbit at the same speed as Earth turns, so they stay in the same spot over Earth's surface.

Hubble Space Telescope
345 miles (555 km)

Polar sun-synchronous satellites
373–497 miles (600–800 km)

1,244 miles (2,000 km)

HIGH EARTH ORBIT ZONE

MEDIUM EARTH ORBIT ZONE

LOW EARTH ORBIT ZONE

GPS satellites
12,600 miles (22,200 km)
Objects orbit once every 12 hours, or twice a day.

International Space Station
255 miles (410 km)

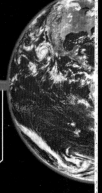

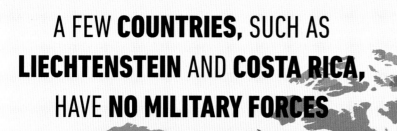

A FEW **COUNTRIES,** SUCH AS **LIECHTENSTEIN** AND **COSTA RICA,** HAVE **NO MILITARY FORCES**

UK

In 2010, the UK spent $56 billion on its armed forces, making it the world's fifth-biggest military spender

USA
The USA spends almost $934 billion per year on its armed forces—more than the next seven-biggest spending countries added together.

France
France holds the world's third-largest nuclear arsenal, with 300 active warheads.

Israel
All Israeli men and women must serve for 2 to 3 years in their armed forces. Israel is the only country to make service for women mandatory.

KEY
The total amount of military expenditure by all the countries of the world in 2010 was $1.83 trillion, which is equivalent to $235 for every person on the planet—almost double what was spent per capita in 2001. The map shows the total number of military vehicles, hardware, and weapons held by selected major countries.

Up to 10 large warships
(including aircraft carriers, cruisers, destroyers, frigates, and corvettes)

Up to 10 submarines

Up to 500 combat-capable aircraft

Up to 1,000 main battle tanks

Up to 500 nuclear warheads

Egypt
All Egyptian men between 18 and 30 must serve in the army for 1 to 3 years.

Brazil
Brazil's armed forces are the largest in South America. The army takes an active role in education, health care, and the construction of roads and railroads.

Armed forces

Sky-high warfare
Armed forces are increasingly using unmanned drones for surveillance or to launch missiles. Drones are controlled remotely from the ground, so air crew is not risked during missions.

Almost all countries have a military—an organized force of soldiers and weapons that defends the country against threats from outside or within. Many countries believe that a large, well-equipped military will discourage others from attacking.

THE US HAS 4 PERCENT OF THE WORLD'S POPULATION BUT

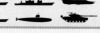

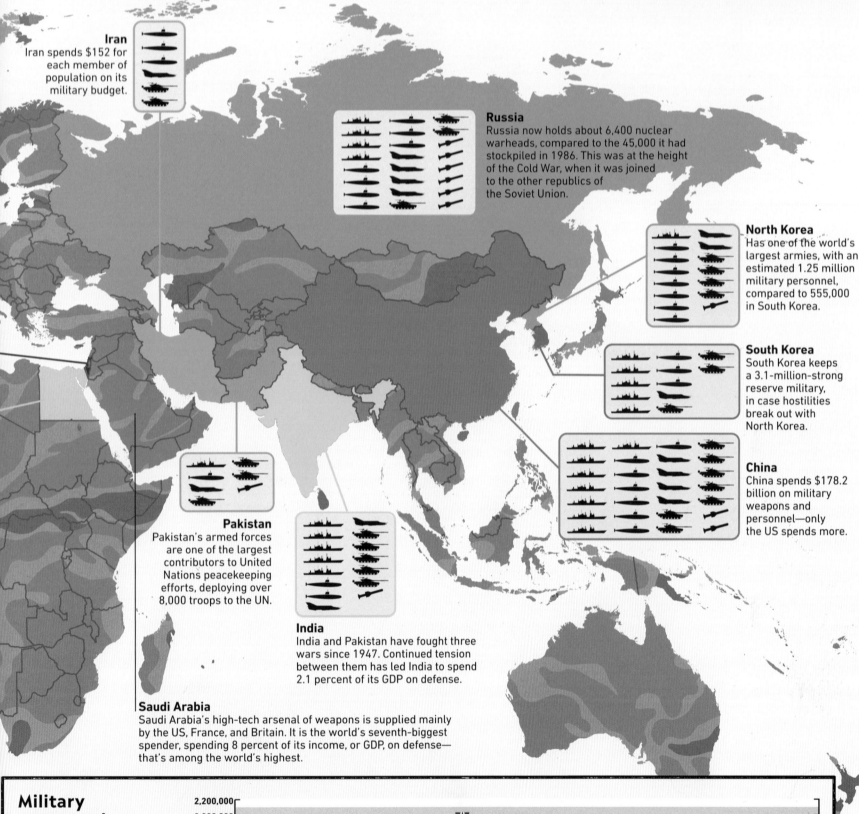

Iran
Iran spends $152 for each member of population on its military budget.

Russia
Russia now holds about 6,400 nuclear warheads, compared to the 45,000 it had stockpiled in 1986. This was at the height of the Cold War, when it was joined to the other republics of the Soviet Union.

North Korea
Has one of the world's largest armies, with an estimated 1.25 million military personnel, compared to 555,000 in South Korea.

South Korea
South Korea keeps a 3.1-million-strong reserve military, in case hostilities break out with North Korea.

China
China spends $178.2 billion on military weapons and personnel—only the US spends more.

Pakistan
Pakistan's armed forces are one of the largest contributors to United Nations peacekeeping efforts, deploying over 8,000 troops to the UN.

India
India and Pakistan have fought three wars since 1947. Continued tension between them has led India to spend 2.1 percent of its GDP on defense.

Saudi Arabia
Saudi Arabia's high-tech arsenal of weapons is supplied mainly by the US, France, and Britain. It is the world's seventh-biggest spender, spending 8 percent of its income, or GDP, on defense— that's among the world's highest.

Military personnel

China commands the world's largest active military force of more than 2 million—but this is only one-and-a-half soldiers in every thousand people. In North Korea, a massive one-fifth of males ages 17–54 are in the regular armed forces.

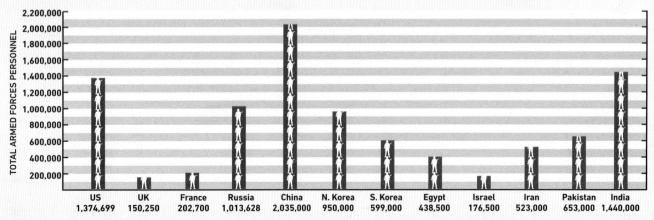

US 1,374,699 · UK 150,250 · France 202,700 · Russia 1,013,628 · China 2,035,000 · N. Korea 950,000 · S. Korea 599,000 · Egypt 438,500 · Israel 176,500 · Iran 523,000 · Pakistan 653,000 · India 1,440,000

History

Easter Island statues
The giant statues, or *moai*, on this small Pacific island stand up to 33 ft (10 m) tall. They were carved with stone tools, mainly between 1250 and 1500, by the Polynesian people who settled the island.

Introduction

Human history is crammed full of incidents, from civilizations rising and falling as wars are fought and lost, to revolutions sweeping away the past to begin again. There has also been great architecture and many important innovations, from the first stone tools that enabled people to hunt animals to radio telescopes that can "see" into deep space.

The Great Sphinx
This statue at Giza, in Egypt, has a human head on a lion's body. It is thought to have been made about 4,500 years ago.

c.200,000 years ago
Modern humans
The *Homo sapiens* species (modern humans) evolves in east Africa.

c.2.4 million years ago
Earliest tools
The first stone tools are made by *Homo habilis*, an early human species.

c.100,000 years ago
Jewelry
Early people wear jewelry made from shell beads.

1227
Genghis Khan
At the death of its Mongol leader Genghis Khan, the Mongol Empire stretches across northern Asia.

1095–1272
The Crusades
Christian and Muslim armies fight nine wars to control Jerusalem.

1200
Holy Roman Empire
This "superpower" of the Middle Ages covers much of central Europe.

900
Khmer dominance, Asia
With their capital at Angkor, the Khmers rule over a large part of Southeast Asia.

1235
Battle of Kirina, Africa
Mandinka forces defeat the Sosso, leading to the birth of the Mali Empire.

1325
Templo Mayor, Mexico
Human sacrifices are made at this temple in the Aztec capital city of Tenochtitlan.

1300
Kanem Empire, Africa
Located north of Lake Chad, Kanem grow powerful and wealthy through its control of trade.

1350
Kingdom of Zimbabwe
The capital of this southern African kingdom is Great Zimbabwe, a stone-walled city.

1949
Chinese Revolution
Led by Mao Zedong, Chinese Communists take power after a long civil war.

1947
Indian independence
After a largely nonviolent rebellion, India wins its independence from Britain.

1945–54
First Indochina War
Indochina (Vietnam, Cambodia, and Laos) wins independence from France.

1939–45
World War II
Allied forces (Britain, France, the US, USSR, and others) at war with Germany, Japan, and Italy.

1950–53
Korean War
Civil war: China and the USSR help North Korea, the United Nations helps South Korea.

1965
Indo-Pakistani War
Conflict between India and Pakistan over the disputed region of Kashmir.

1955–75
Vietnam War
Communist North Vietnam triumphs over South Vietnam, which is aided by US forces.

1969
Concorde
The world's first supersonic airliner, Concorde, flies for the first time.

2011
World's longest bridge, China
Completion of the 102.4-mile- (164.8-km-) long Danyang-Kunshan Grand Bridge.

2020
COVID-19
Outbreak of a newly discovered coronavirus causes a global pandemic, with up to 2.6 million deaths in the first year.

2011
"Arab Spring"
Revolution and protest sweep through Egypt, Libya, and other Arab countries.

c.90,000 years ago
Burial rites
People begin burying their dead along with meaningful objects such as beads.

c.3200 BCE
Pirámide Mayor, Peru
Built by the Norte Chico civilization at Caral, the most ancient city in the Americas.

1450 BCE
New Kingdom of Egypt
Egypt's empire stretches north to Syria and south to Nubia (modern Sudan).

490 BCE
First Persian Empire
Persia rules territory from the edge of India to Egypt and Greece, linking East with West.

265 BCE
Mauryan Empire, Asia
Under Ashoka, the Mauryan Empire extends over almost all of the Indian subcontinent.

c.40,000 years ago
First music and art
Music is played on simple flutes, and figurines are carved from stone.

c.2589–2500 BCE
Pyramids of Giza, Egypt
Vast tombs are built for the Egyptian pharaohs Khufu, Khafre, and Menkaure.

c.700 BCE
Olmec civilization
Mexico's Olmec culture reaches its peak. It will influence the later Mayan and Aztec cultures.

323 BCE
Macedonian Empire
King Alexander the Great of Macedonia rules lands from Greece to the edge of India.

264–146 BCE
Punic Wars
Three wars erupt between Rome and Carthage, North Africa. Rome emerges victorious.

750
Umayyad Caliphate
The second of four great Islamic dynasties, with its capital in Damascus (Syria).

650
Huari Empire, Peru
The highly organized Huari, in Peru, conquer and control much of the Andean region.

c.300 CE
Mayan culture, Central America
Established by 1000 BCE, Mayan civilization is now at its height. It will last until 1697 CE.

100 CE
Pyramid of the Sun, Mexico
One of two huge stepped pyramids is built in the city of Teotihuacán.

87 BCE
Han Dynasty, China
A time of prosperity in China and an expansion of territories ruled by China.

700
Tihuanaco, Peru/Bolivia
This strong state is centered on a bustling city beside Lake Titicaca in the Andes.

555
Byzantine power
Byzantine rule extends over North Africa and the eastern part of the old Roman Empire.

117 CE
Roman supremacy
Rome now controls much of Europe, north Africa, and the Middle East.

80 CE
Colosseum, Rome
Opening of the stadium in Rome where gladiators fought to the death.

214 BCE
Great Wall of China
Construction begins of this vast defensive wall along China's northern border.

Colosseum, Rome

1453
Fall of Constantinople
The capital of the Byzantine Empire falls to invading Muslim Ottoman forces.

1500
Songhai power, Africa
The Songhai control the Niger Valley, west to Senegal and east to Agades (modern Niger).

1532
Battle of Cajamarca, Peru
Spanish invaders defeat the Inca forces of Atahualpa, leading to 300 years of Spanish rule.

1683
Battle of Vienna
Ottoman expansion finally halts with a defeat by the Holy Roman Empire.

1450
Machu Picchu, Peru
A secret hilltop city of the Incas, who will dominate northern South America.

1500
Ming Dynasty, China
After throwing out the Mongols, China restores its culture and expands its borders.

1519
Aztec rule, Mexico
The Aztecs now rule more than 25 million people. In 1521, they are conquered by the Spanish.

1642–51
English Civil War
Parliamentarians defeat Royalists, leading to the execution of King Charles I.

1690
Mughal Empire, India
Under Aurangzeb, the Islamic Mughal Empire of India is at its most powerful.

1922
Height of British Empire
Britain's empire now covers more than 20 percent of the world's land area.

1914–18
World War I
Britain, France, the US, and other allies battle Germany, Austria-Hungary, and Turkey.

1880–1902
Boer Wars, Africa
Two wars are fought between Dutch Boer settlers in South Africa and Britain.

1819–30
South American independence
Independence from Spain for Colombia, Peru, Bolivia, Ecuador, and Venezuela.

1789–99
French Revolution
Overthrow of the French monarchy in a bloody revolution. France becomes a republic.

1917
Russian Revolution
Revolt against rule by Tsar Nicholas II; Russia becomes Communist.

1912
Sinking of the *Titanic*
More than 1,500 people die when this luxury liner hits an iceberg and sinks.

1861–65
American Civil War
War between the southern Confederate states and the Union states of the north.

1799–1815
Napoleonic era
France, led by Napoleon Bonaparte, is the dominant military power in Europe.

1775–83
American Revolutionary War
With the help of France and other countries, the US wins independence from Britain.

1980
Very Large Array
In New Mexico, this giant radio astronomy observatory is completed.

Sydney Opera House
Opened in 1973, this arts venue in Sydney, Australia, was designed by Danish architect Jørn Utzon.

1994
End of Apartheid
South Africa's official segregation policy, Apartheid, ends and equality is reached for Black South Africans.

1989–1991
End of Communist bloc
Communist regimes in many countries of eastern Europe are overthrown.

ONLY VIRUS TO BE ERADICATED THROUGH VACCINATION.

Australopithecus

Australopithecus hominins evolved about 4.2 million years ago in east Africa. Six species are known. One species, called *A. afarensis*, may be the ancestor of humans. Fossils show that it was up to 5 ft (1.5 m) tall and had a relatively small brain. Crucially, it could walk upright.

Paranthropus

The three *Paranthropus* species had a bony crest on top of the skull to anchor strong chewing muscles. *P. boisei* is nicknamed "nutcracker man" because of its massive jaws and cheek teeth.

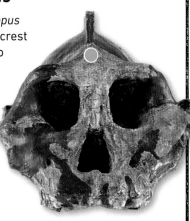

Neander Valley, Germany
A partial skeleton of *H. neanderthalensis* found in a cave here in 1856 was the first fossil to be identified as human remains.

Fossil humans

Fossil discoveries have helped scientists to piece together the story of human evolution. Modern humans—*Homo sapiens*—and their ancestors are called hominins. *Sahelanthropus tchadensis*, the first hominin, was an apelike animal that appeared in Africa about 7 million years ago. Later hominin species left Africa and spread out around the world.

Laetoli, Tanzania
Footprints of at least two *Australopithecus afarensis* individuals were discovered here, preserved in volcanic ash.

Olduvai Gorge, Tanzania
Stone tools and fossils of *P. boisei* and *H. habilis* were found here.

South Africa
Finds include *Australopithecus*, *Paranthropus*, *H. habilis*, and *H. sapiens* fossils.

Apelike
Australopithecus—
six species

4 million years ago

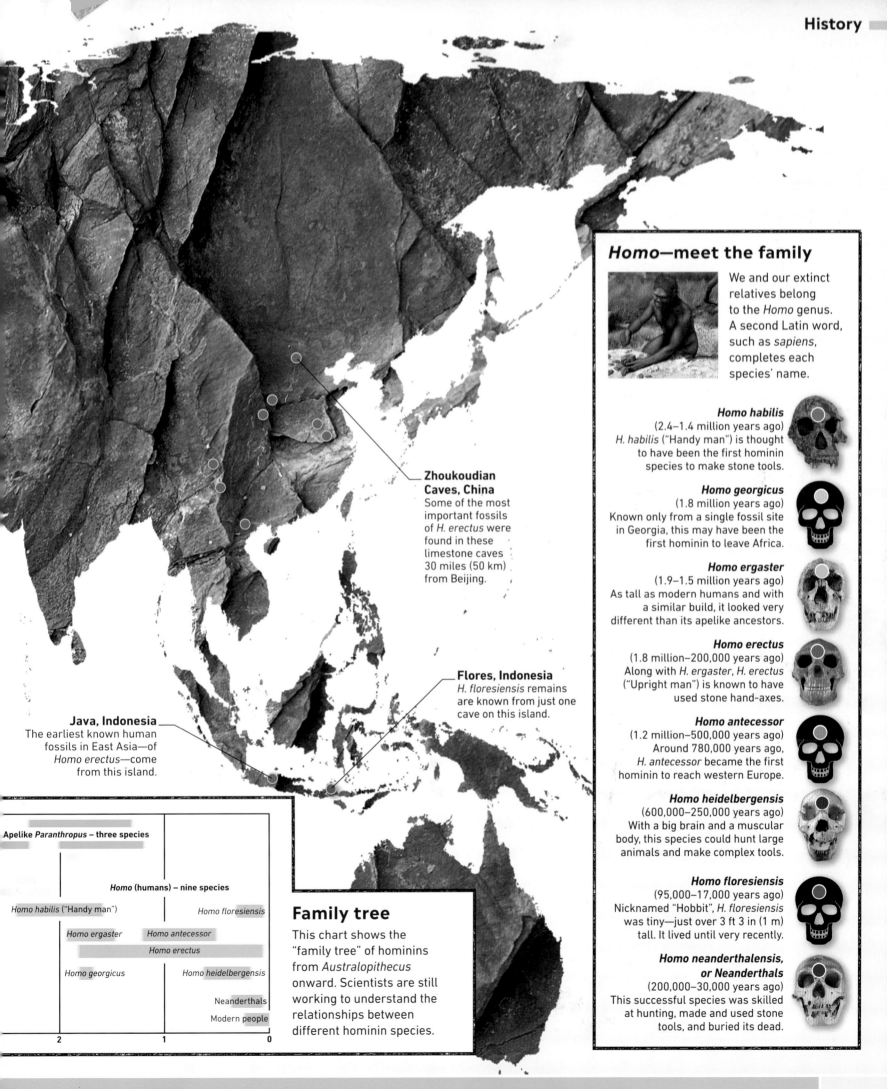

Zhoukoudian Caves, China
Some of the most important fossils of *H. erectus* were found in these limestone caves 30 miles (50 km) from Beijing.

Flores, Indonesia
H. floresiensis remains are known from just one cave on this island.

Java, Indonesia
The earliest known human fossils in East Asia—of *Homo erectus*—come from this island.

Homo—meet the family

We and our extinct relatives belong to the *Homo* genus. A second Latin word, such as *sapiens*, completes each species' name.

Homo habilis
(2.4–1.4 million years ago)
H. habilis ("Handy man") is thought to have been the first hominin species to make stone tools.

Homo georgicus
(1.8 million years ago)
Known only from a single fossil site in Georgia, this may have been the first hominin to leave Africa.

Homo ergaster
(1.9–1.5 million years ago)
As tall as modern humans and with a similar build, it looked very different than its apelike ancestors.

Homo erectus
(1.8 million–200,000 years ago)
Along with *H. ergaster*, *H. erectus* ("Upright man") is known to have used stone hand-axes.

Homo antecessor
(1.2 million–500,000 years ago)
Around 780,000 years ago, *H. antecessor* became the first hominin to reach western Europe.

Homo heidelbergensis
(600,000–250,000 years ago)
With a big brain and a muscular body, this species could hunt large animals and make complex tools.

Homo floresiensis
(95,000–17,000 years ago)
Nicknamed "Hobbit", *H. floresiensis* was tiny—just over 3 ft 3 in (1 m) tall. It lived until very recently.

**Homo neanderthalensis,
or Neanderthals**
(200,000–30,000 years ago)
This successful species was skilled at hunting, made and used stone tools, and buried its dead.

Family tree

This chart shows the "family tree" of hominins from *Australopithecus* onward. Scientists are still working to understand the relationships between different hominin species.

Apelike *Paranthropus* – three species

Homo (humans) – nine species

Homo habilis ("Handy man")		*Homo floresiensis*
Homo ergaster	*Homo antecessor*	
	Homo erectus	
Homo georgicus		*Homo heidelbergensis*
		Neanderthals
		Modern people
2	1	0

Prehistoric culture

Music, art, religion, and technology all began so long ago, we can't be certain of exactly when. There are clues to early culture, however, such as ritual burial sites, which archaeologists can date.

Earliest music

Music, like art, is much older than writing, since bone flutes and other musical instruments have been made and played for more than 40,000 years.

◆ Early instrument site

Antler flute, Hohle Fels, Germany, 43,000 years ago

First jewelry

People wore jewelry more than 100,000 years ago in sites as distant as Israel and South Africa.

◆ Early jewelry site

Shell beads, Balzi Rossi, Italy

Changes in stone tools

2.4 million years ago
The earliest tools, called the Oldowan tool kit, were made by an early human species called "Handy man," or *homo habilis*, in Africa. Oldowan-style tools in Europe and Asia are much younger, made by later types of humans, including Neanderthals.

● Oldowan site

1.8 million years ago
The Acheulean tool kit of our later ancestors, such as *Homo erectus*, included a new invention—the hand ax, with a finely chiseled edge.

● Acheulean site

200,000 years ago
Mousterian tools spanned the Middle Stone Age (ended around 40,000 BCE) and included lots of specialized shapes for different jobs.

● Mousterian site

13,000 years ago
The earliest stone tools discovered in America are from the 13,000-year-old "Clovis" people.

● Clovis site

East Wenatchee, Washington, US

Walker, Minnesota, US

Horseshoe Canyon paintings, Utah, US

Cactus Hill, Virginia, US

Clovis, New Mexico, US

Salado, Texas, US

Wicklow Pipes, Ireland

Shell bead necklace, Cro-Magnon, France

Lascaux Caves, France

Altamira and El Castillo caves, Spain. El Castillo features the oldest known paintings, made 40,800 years ago, possibly by Neanderthals

Lady of Brassempouy carving, France

Ivory horse figurine, Lourdes, France

Shell beads, Grotte des Pidgeons, Morocco

Algerian Sahara

Serra de Capivara paintings, Brazil

Cueva de las Manos paintings, Argentina

Cueva del Milodon, Chile

THE FIRST KNOWN SEWING NEEDLE DATES BACK ABOUT 25,000 YEARS

Flutes, Hohle Fels Cave, and Geissenklösterle, Germany

Bisovava, Russia

"Lion Man" bone carving, Germany

Sungir Graves, Russia

Carved ivory running lion, Czech Republic

Tata Plaque (mysterious object made by a Neanderthal 100,000 years ago), Hungary

Krapina, Croatia

Tbilisi, Georgia

Balzi Rossi caves, Italy

Pechka rock shelter, Armenia

Shanidar Cave, Iraq

Kashafrud, Iran

Riwat, Pakistan

Qafzeh, Israel

Chauvet Cave, France

Shell beads, Skhūl, Israel

Gebelein, Egypt

Gona, Ethiopia (world's oldest tools)

Konso-Gardula, Ethiopia

Omo, Ethiopia

Turkana, Kenya

Lokalalei, Kenya

Olduvai Gorge, Tanzania

Twin Rivers, Zambia

Inanke Cave, Zimbabwe

Sterkfontein, South Africa

"Apollo 11" rock shelter, Namibia

Swartkrans, South Africa

Shell beads, Blombos Cave, South Africa

Majuangou, China

Carved bone disk, Xiaogushan, China

Bone flutes, Jiahu, China

Bhimbetka paintings, India

Bose, China

Isampur, India

Island of Socotra, Yemen

Kakadu National Park, Australia

THE **OLDEST KNOWN CLAY POTS** WERE MADE IN **CHINA** ABOUT **20,000** YEARS AGO

Earliest burials

Our ancestors began burying their dead with significant objects, such as beads or other decorations, at least 100,000 years ago.

◆ Early burial site

Skull with shells, 25,000 years ago, Balzi Rossi, Italy

Earliest paintings

Humans have painted and carved rock surfaces since at least 40,000 years ago. Some paintings show people dancing and singing.

◆ Early painting site

Inanke Cave, Zimbabwe, 5,000–10,000 years ago

The first sculpture

The earliest known sculpture consists of figurines carved from stone and bone to look like humans and animals. Some date back up to 40,000 years.

◇ Site of artwork

"Lion Man," Germany, 40,000 years ago

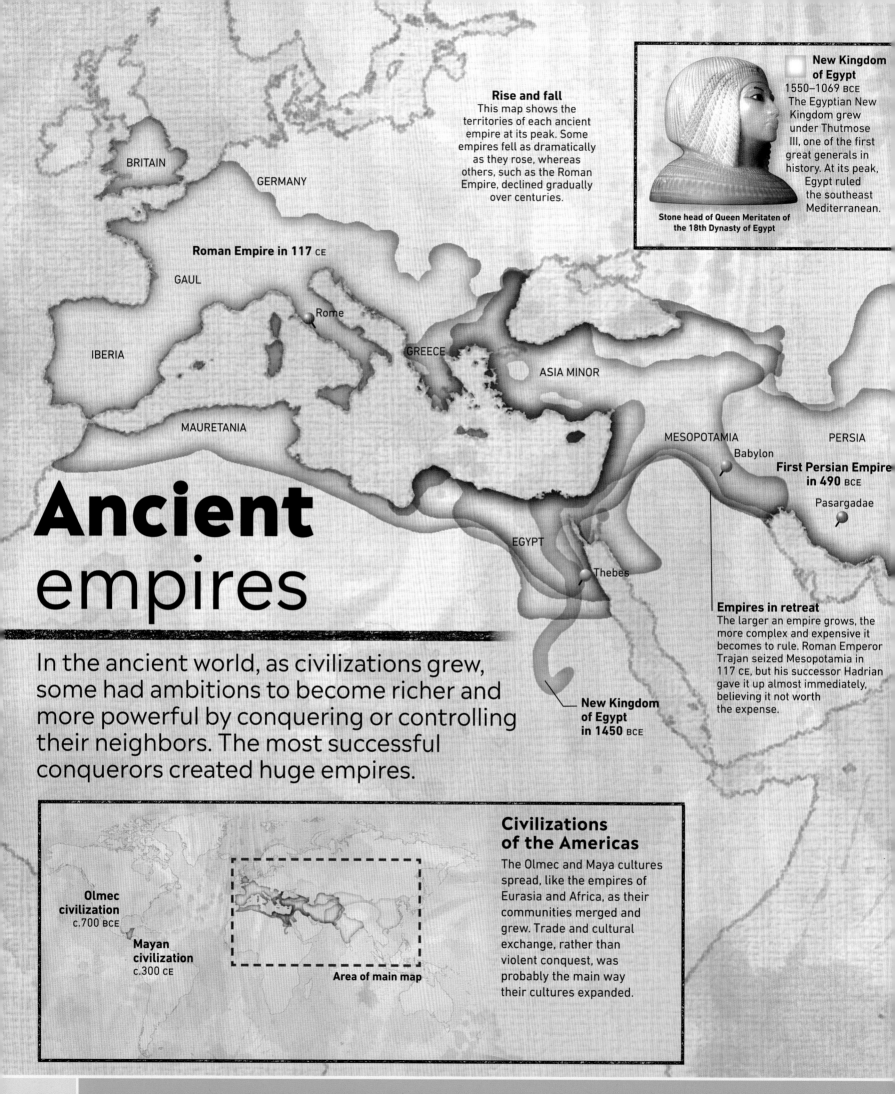

Ancient empires

In the ancient world, as civilizations grew, some had ambitions to become richer and more powerful by conquering or controlling their neighbors. The most successful conquerors created huge empires.

Rise and fall
This map shows the territories of each ancient empire at its peak. Some empires fell as dramatically as they rose, whereas others, such as the Roman Empire, declined gradually over centuries.

BRITAIN
GERMANY

Roman Empire in 117 CE

GAUL
Rome
IBERIA
GREECE
ASIA MINOR
MAURETANIA
MESOPOTAMIA
PERSIA
Babylon
First Persian Empire in 490 BCE
Pasargadae
EGYPT
Thebes

New Kingdom of Egypt in 1450 BCE

New Kingdom of Egypt
1550–1069 BCE
The Egyptian New Kingdom grew under Thutmose III, one of the first great generals in history. At its peak, Egypt ruled the southeast Mediterranean.

Stone head of Queen Meritaten of the 18th Dynasty of Egypt

Empires in retreat
The larger an empire grows, the more complex and expensive it becomes to rule. Roman Emperor Trajan seized Mesopotamia in 117 CE, but his successor Hadrian gave it up almost immediately, believing it not worth the expense.

Civilizations of the Americas
The Olmec and Maya cultures spread, like the empires of Eurasia and Africa, as their communities merged and grew. Trade and cultural exchange, rather than violent conquest, was probably the main way their cultures expanded.

Olmec civilization
c.700 BCE

Mayan civilization
c.300 CE

Area of main map

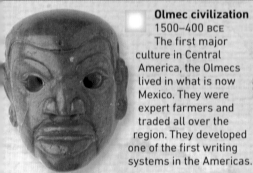

Olmec civilization
1500–400 BCE
The first major culture in Central America, the Olmecs lived in what is now Mexico. They were expert farmers and traded all over the region. They developed one of the first writing systems in the Americas.

Olmec stone mask

First Persian Empire
550–336 BCE
Cyrus the Great and his army conquered huge swathes of central Asia and grabbed enormous wealth from the kingdoms they conquered. Cyrus's successor, Darius I, built cities, roads, and even a canal from the Nile river to the Red Sea.

Ornate Persian silver bowl

Empire of Alexander the Great
330–323 BCE
Alexander was a general from Macedon, a kingdom north of Greece. At its height, his empire covered most of the world known to Greeks. For centuries after his death, the Greek culture that he introduced continued to dominate the eastern Mediterranean and western Asia.

Coin showing Alexander the Great's head

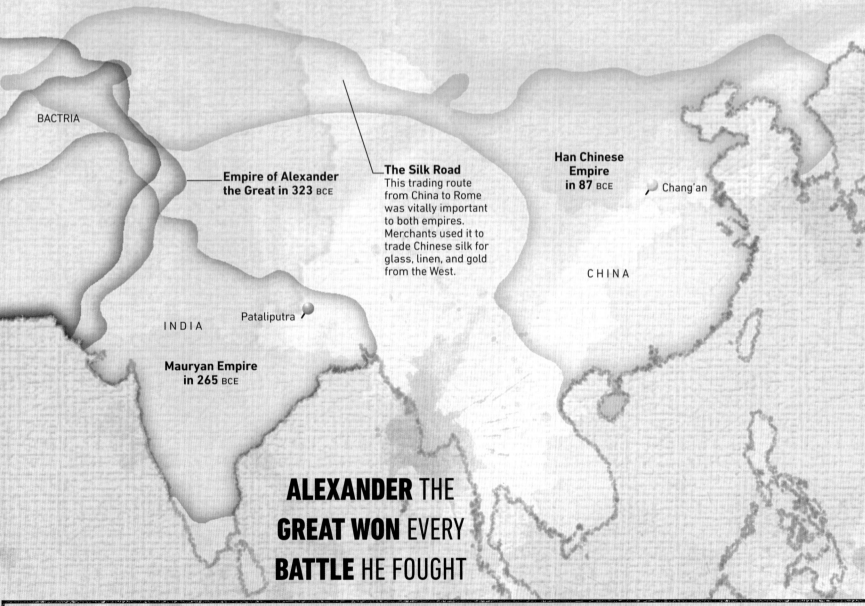

BACTRIA

Empire of Alexander the Great in 323 BCE

The Silk Road
This trading route from China to Rome was vitally important to both empires. Merchants used it to trade Chinese silk for glass, linen, and gold from the West.

Han Chinese Empire in 87 BCE

Chang'an

CHINA

Pataliputra

INDIA

Mauryan Empire in 265 BCE

ALEXANDER THE GREAT WON EVERY BATTLE HE FOUGHT

Mauryan Empire
321–185 BCE
Chandragupta Maurya was the first leader to conquer the entire Indian subcontinent. His son Ashoka became a Buddhist and ruled the empire peacefully for 42 years.

Mauryan figure

Han Empire
206–220 CE
The four centuries of Han rule are often called the Golden Age of Ancient China. It was an era of peace and prosperity in which China became a major world power.

Han pot

Roman Empire
27 BCE–476 CE
One of history's most influential civilizations, Rome controlled much of Europe, western Asia, and north Africa. Many roads, aqueducts, and canals built by the Romans are still in use today.

Head of Emperor Claudius

Mayan civilization
500–900 CE
One of the most advanced cultures of the ancient world, the Maya developed an accurate yearly calendar based on their sophisticated understanding of astronomy.

Mayan statuette

Ancient wonders

Ancient Greek travelers and authors such as Herodotus, Antipater, and Philo of Byzantium praised the architectural marvels of the age in their writings. The buildings and statues they described became known as the "Seven Wonders of the World." Today, we recognize many other amazing structures that architects, masons, and sculptors of the past built with relatively simple tools.

Stonehenge
c.2600 bce,
Wiltshire, England

Carnac
c.3300 BCE, Brittany, France.
Stone Age monument
of more than 3,000
standing stones.

Pont-du-Gard
19 BCE, Nîmes, France

Colosseum
80 CE, Rome, Italy

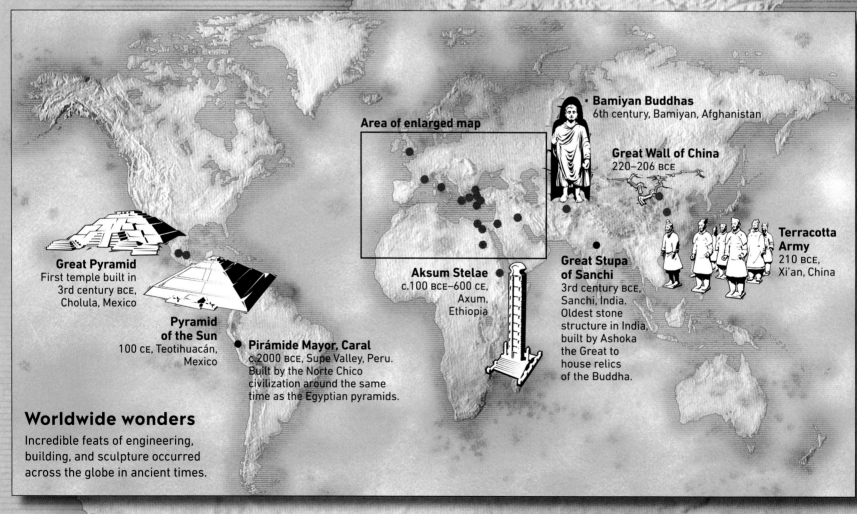

Bamiyan Buddhas
6th century, Bamiyan, Afghanistan

Great Wall of China
220–206 BCE

Area of enlarged map

Great Pyramid
First temple built in
3rd century BCE,
Cholula, Mexico

**Pyramid
of the Sun**
100 CE, Teotihuacán,
Mexico

Pirámide Mayor, Caral
c.2000 BCE, Supe Valley, Peru.
Built by the Norte Chico
civilization around the same
time as the Egyptian pyramids.

Aksum Stelae
c.100 BCE–600 CE,
Axum,
Ethiopia

**Great Stupa
of Sanchi**
3rd century BCE,
Sanchi, India.
Oldest stone
structure in India,
built by Ashoka
the Great to
house relics
of the Buddha.

**Terracotta
Army**
210 BCE,
Xi'an, China

Worldwide wonders

Incredible feats of engineering,
building, and sculpture occurred
across the globe in ancient times.

THE COLOSSEUM IN ROME COULD HOLD MORE THAN 50,000 SPECTATORS

Seven Wonders of the World

Only the pyramids at Giza still stand. Earthquakes destroyed the Hanging Gardens, the Colossus, and the Pharos; flooding and fire ruined the Mausoleum and the Statue of Zeus. The Temple of Artemis was wrecked by the Goths.

Pyramids of Giza
Built as tombs for the pharaohs Khufu, Khafre, and Menkaure.

Hanging Gardens of Babylon
Nebuchadnezzar II built these lush, terraced gardens for his wife, Amytis.

Mausoleum at Halicarnassus
Tomb of Persian governor Mausolus, famed for its size and lavish carvings.

Temple of Artemis
Dedicated to the Greek goddess of hunting, chastity, and childbirth.

Colossus of Rhodes
Vast bronze-and-iron statue, 105 ft (32 m) tall, of the Greek sun-god Helios.

Pharos of Alexandria
A fire at the top of this huge lighthouse was visible from 30 miles (50 km) away.

Statue of Zeus in Olympia
The sculptor Phidias built this 43-ft (13-m) statue of the king of the gods.

Other ancient wonders

These wonders didn't make the Seven Wonders list, mainly because they were unknown to the Greeks. Some of them were built during later periods.

Colosseum
Stadium where gladiators fought to the death.

Hagia Sofia
Enormous, richly decorated church, later a mosque.

Petra
A city hewn out of rock. Capital of the Nabataeans.

Temples of Abu-Simbel
Two temples built to honor the pharaoh Rameses II.

Pont-du-Gard
Roman aqueduct that carried water to Nîmes.

Acropolis
Greek citadel that includes the Parthenon Temple.

Great Pyramid
World's largest pyramid, now with a church on top.

Pyramid of the Sun
Steep steps up the side led to a temple on the top.

Stonehenge
Prehistoric monument with a circle of enormous stones.

Bamiyan Buddhas
Huge statues chiseled into a cliff; destroyed in 2001.

Great Wall of China
Once ran for 3,889 miles (6,259 km) along China's northern border.

Terracotta Army
8,000 life-size warriors entombed with the first emperor of China.

Aksum Stelae
A group of memorial obelisks carved from huge blocks of stone.

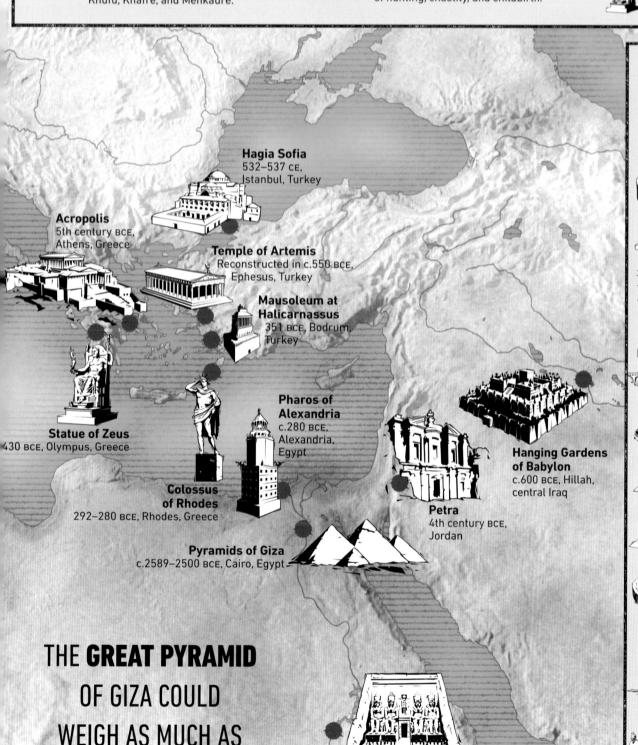

Hagia Sofia
532–537 CE, Istanbul, Turkey

Acropolis
5th century BCE, Athens, Greece

Temple of Artemis
Reconstructed in c.550 BCE, Ephesus, Turkey

Mausoleum at Halicarnassus
351 BCE, Bodrum, Turkey

Statue of Zeus
430 BCE, Olympus, Greece

Colossus of Rhodes
292–280 BCE, Rhodes, Greece

Pharos of Alexandria
c.280 BCE, Alexandria, Egypt

Hanging Gardens of Babylon
c.600 BCE, Hillah, central Iraq

Petra
4th century BCE, Jordan

Pyramids of Giza
c.2589–2500 BCE, Cairo, Egypt

Temples of Abu-Simbel
c.1257 BCE, Abu-Simbel, Egypt

THE **GREAT PYRAMID** OF GIZA COULD WEIGH AS MUCH AS **716** MILLION TONS

Famous mummies

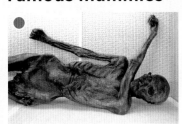

Ötzi the Iceman
About 5,300 years ago, a traveler died when caught in a snowstorm in the Alps. His body became buried in the snow and then froze. In 1991, the corpse was discovered on top of a glacier.

Pharaoh Tutankhamun
The mummy of Tutankhamun was found in a tomb in the Valley of the Kings in 1922. It wore a gold mask and lay inside a nest of three gold cases. The tomb, which had been sealed for 3,200 years, contained statues, furniture, and jewelry.

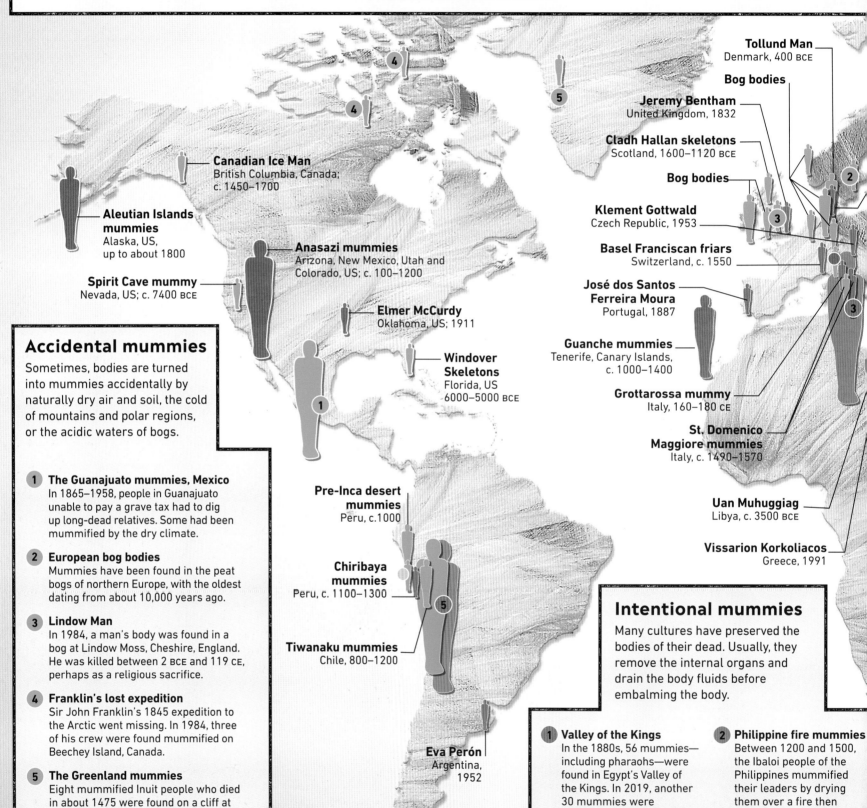

Tollund Man
Denmark, 400 BCE

Bog bodies

Jeremy Bentham
United Kingdom, 1832

Cladh Hallan skeletons
Scotland, 1600–1120 BCE

Bog bodies

Klement Gottwald
Czech Republic, 1953

Basel Franciscan friars
Switzerland, c. 1550

José dos Santos Ferreira Moura
Portugal, 1887

Guanche mummies
Tenerife, Canary Islands, c. 1000–1400

Grottarossa mummy
Italy, 160–180 CE

St. Domenico Maggiore mummies
Italy, c. 1490–1570

Uan Muhuggiag
Libya, c. 3500 BCE

Vissarion Korkoliacos
Greece, 1991

Canadian Ice Man
British Columbia, Canada; c. 1450–1700

Aleutian Islands mummies
Alaska, US, up to about 1800

Spirit Cave mummy
Nevada, US; c. 7400 BCE

Anasazi mummies
Arizona, New Mexico, Utah and Colorado, US; c. 100–1200

Elmer McCurdy
Oklahoma, US; 1911

Windover Skeletons
Florida, US 6000–5000 BCE

Pre-Inca desert mummies
Peru, c.1000

Chiribaya mummies
Peru, c. 1100–1300

Tiwanaku mummies
Chile, 800–1200

Eva Perón
Argentina, 1952

Accidental mummies

Sometimes, bodies are turned into mummies accidentally by naturally dry air and soil, the cold of mountains and polar regions, or the acidic waters of bogs.

1 The Guanajuato mummies, Mexico
In 1865–1958, people in Guanajuato unable to pay a grave tax had to dig up long-dead relatives. Some had been mummified by the dry climate.

2 European bog bodies
Mummies have been found in the peat bogs of northern Europe, with the oldest dating from about 10,000 years ago.

3 Lindow Man
In 1984, a man's body was found in a bog at Lindow Moss, Cheshire, England. He was killed between 2 BCE and 119 CE, perhaps as a religious sacrifice.

4 Franklin's lost expedition
Sir John Franklin's 1845 expedition to the Arctic went missing. In 1984, three of his crew were found mummified on Beechey Island, Canada.

5 The Greenland mummies
Eight mummified Inuit people who died in about 1475 were found on a cliff at Nuuk, Greenland, in 1972. Their bodies had freeze-dried.

Intentional mummies

Many cultures have preserved the bodies of their dead. Usually, they remove the internal organs and drain the body fluids before embalming the body.

1 Valley of the Kings
In the 1880s, 56 mummies—including pharaohs—were found in Egypt's Valley of the Kings. In 2019, another 30 mummies were discovered in this area by Egyptian archaeologists.

2 Philippine fire mummies
Between 1200 and 1500, the Ibaloi people of the Philippines mummified their leaders by drying them over a fire then putting them in caves.

Juanita the Ice Maiden
In 1995, an Inca girl aged 11–15 was found on Mount Ampato, Peru. The discoverers named her Juanita, or the "Ice Maiden." She was sacrificed to the gods about 530 years ago. The cold had preserved her skin, organs, blood, and stomach contents.

MUMMY DISCOVERIES WORLDWIDE
Some mummies are discovered singly, often in remote locations such as in peat bogs or on high mountains. Other finds involve larger numbers of mummies—for example, in communal graves, tombs, caves, or catacombs.

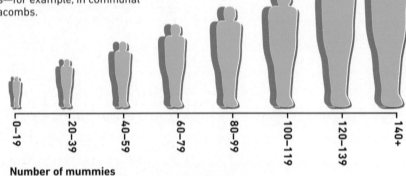

Accidental mummies

Intentional mummies

Number of mummies

0–19 20–39 40–59 60–79 80–99 100–119 120–139 140+

James Hepburn, 4th Earl of Bothwell
Denmark, 1578

Charles Eugène de Croy
Estonia, 1702

Vladimir Lenin
Russia, 1924

Georgi Dimitrov
Bulgaria, 1949

Valley of the Golden Mummies Egypt, 332 BCE–395 CE

...bnitz Girl
...nd,
... BCE

Maronite mummies
Lebanon, 1283

Chehrabad Salt Mine mummies Iran, 4th century BCE–4th century CE

Iufaa and family
Egypt, c.500 BCE

Saqqara mummies
Egypt, 640 BCE

Nubian mummies
Sudan, 250–1400

Tarim mummies
China, 1800–200 BCE

Siberian Ice Maiden
Russia, c. 400 BCE

Pazyryk ice mummies
Mongolia, c. 700–200 BCE

Mao Zedong
China, 1976

Xin Zhui
China, c. 150 BCE

Ho Chi Minh
Vietnam, 1969

Mummy monk "Luang Phor Daeng"
Thailand, c. 1985

Vu Khac Minh and Vu Khac Truong
Vietnam, c. 1600–1700

Chiang Kai-shek and Chiang Ching-kuo
Taiwan, 1975 and 1988

Buddhist self-mummified nun and monks
Taiwan, 1680–1830,

Fujiwara clan mummies
Japan, 1128–1189

Kim Il-Sung and Kim Jong-il
North Korea, 1994 and 2011

Korean mummies
South Korea, c. 1350–1500

Lost mummies of New Guinea
Papua New Guinea, up to 1950s

Mummies

Mummies—the preserved bodies of the dead—have been found the world over. Many were made deliberately, while others formed naturally. More recently, some countries have mummified their leaders.

3 Mummies of Palermo
In 1599, Christian monks in Palermo, Sicily, began to mummify their dead and stored them in catacombs. Later, rich people paid the monks to mummify their bodies.

4 Self-mummified monks
From 1680–1830, some Buddhist monks in Japan mummified themselves. They starved, drank special tea to make their body toxic to maggots, and then were sealed alive in a stone tomb.

5 Chinchorro mummies
The Chinchorro, who lived in what is now Chile and Peru, were the first people known to make mummies. Their oldest mummies date from as early as 5000 BCE.

THE PALERMO **CATACOMBS** CONTAIN ABOUT **8,000 MUMMIES**

North America

The Maya and Aztecs built spectacular pyramid-temples. Human sacrifice took place on the Templo Mayor in the Aztec capital Tenochtitlán (now Mexico City).

Templo Mayor, Mexico

Notre Dame de Paris
Paris, France, 1163–1345

Angel Mounds
Evansville, Indiana, 1000 CE

St. Paul's Cathedral (first building)
London, England, 604 CE

Cahokia Mounds and Monks Mound
Collinsville, Illinois, 600–1400 CE

Parkin Indian Mound
Parkin, Arkansas, 1350

Cluny Abbey
Burgundy, France, 910 CE

Taos Pueblo
New Mexico, between late 1200s and mid-1500s

Kincaid Mounds
Brookport, Illinois, 1050–1400 CE

Great Houses of the Chacoan people,
Chaco Canyon, New Mexico, between 900 CE and 1150 CE

Moundville settlement
Alabama, 1000–1450 CE

Alhambra
Granada, Spain, 14th century

Templo Mayor
Mexico City, Mexico, first built 1325, rebuilt six times

Ocmulgee Great Temple Mound
Macon, Georgia, 950–1150 CE

Benin Bronzes
Kingdom of Benin (in modern Nigeria), 13th–16th century

Timbuktu
Mali, 12th century

El Castillo
Chichen Itza, Mexico, 9th–12th century

Calixtlahuaca
Toluca, Mexico, 1100–1520 CE

Temple of the Inscriptions
Chiapas, Mexico, 683 CE

Cusco and the Koricancha
Vilcabamba, Cusco, Peru, between 1200s and 1532

Machu Picchu
Vilcabamba, Cuzco, Peru, 1450

Isla del Sol
Lake Titicaca, Bolivia, 15th century

Royal Palaces of Abomey
Dahomey (modern Benin), 1695

Medieval wonders

El Fuerte de Samaipata
Bolivia, 14th century

Sacsayhuamán
Cuzco, Peru, between early 1400s and mid-1500s

South America

By the late 1400s, the Incas had a vast empire in western South America. The city of Machu Picchu occupied a remote hilltop at the edge of the empire.

Machu Picchu

Ollantaytambo
Cuzco, Peru, mid-15th century

"Medieval" means the Middle Ages, which lasted from the 5th century to the end of the 15th century. The period ended when the world became connected by explorers such as Columbus, heralding the start of modern times. Medieval times saw amazing architectural feats worldwide.

Moai figures
Easter Island, Chile, between 1100 CE and 1650 CE

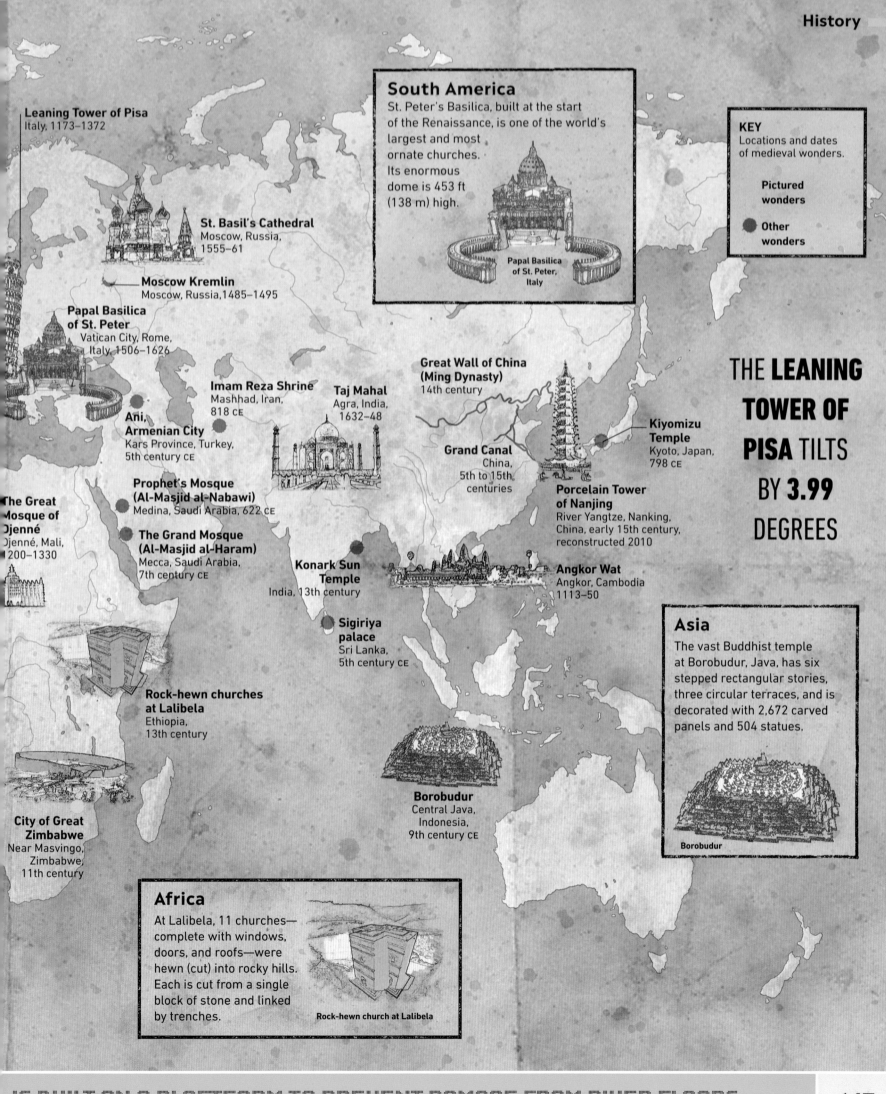

Leaning Tower of Pisa
Italy, 1173–1372

South America
St. Peter's Basilica, built at the start of the Renaissance, is one of the world's largest and most ornate churches. Its enormous dome is 453 ft (138 m) high.

Papal Basilica of St. Peter, Italy

KEY
Locations and dates of medieval wonders.

Pictured wonders

Other wonders

St. Basil's Cathedral
Moscow, Russia, 1555–61

Moscow Kremlin
Moscow, Russia,1485–1495

Papal Basilica of St. Peter
Vatican City, Rome, Italy, 1506–1626

Great Wall of China (Ming Dynasty)
14th century

Imam Reza Shrine
Mashhad, Iran, 818 CE

Taj Mahal
Agra, India, 1632–48

Kiyomizu Temple
Kyoto, Japan, 798 CE

THE **LEANING TOWER OF PISA** TILTS BY **3.99** DEGREES

Ani, Armenian City
Kars Province, Turkey, 5th century CE

Grand Canal
China, 5th to 15th centuries

The Great Mosque of Djenné
Djenné, Mali, 1200–1330

Prophet's Mosque (Al-Masjid al-Nabawi)
Medina, Saudi Arabia, 622 CE

Porcelain Tower of Nanjing
River Yangtze, Nanking, China, early 15th century, reconstructed 2010

The Grand Mosque (Al-Masjid al-Haram)
Mecca, Saudi Arabia, 7th century CE

Konark Sun Temple
India, 13th century

Angkor Wat
Angkor, Cambodia 1113–50

Asia
The vast Buddhist temple at Borobudur, Java, has six stepped rectangular stories, three circular terraces, and is decorated with 2,672 carved panels and 504 statues.

Sigiriya palace
Sri Lanka, 5th century CE

Rock-hewn churches at Lalibela
Ethiopia, 13th century

Borobudur
Central Java, Indonesia, 9th century CE

Borobudur

City of Great Zimbabwe
Near Masvingo, Zimbabwe, 11th century

Africa
At Lalibela, 11 churches—complete with windows, doors, and roofs—were hewn (cut) into rocky hills. Each is cut from a single block of stone and linked by trenches.

Rock-hewn church at Lalibela

IS BUILT ON A PLAFTFORM TO PREVENT DAMAGE FROM RIVER FLOODS.

Medieval empires

At times between 500 and 1500 CE, one power or another controlled vast parts of Europe and Asia, and spread Islam and Christianity across the world as they knew it. Little known to them, African rulers joined up large regions for the first time, while empires in the Americas grew in isolation from the rest of the world.

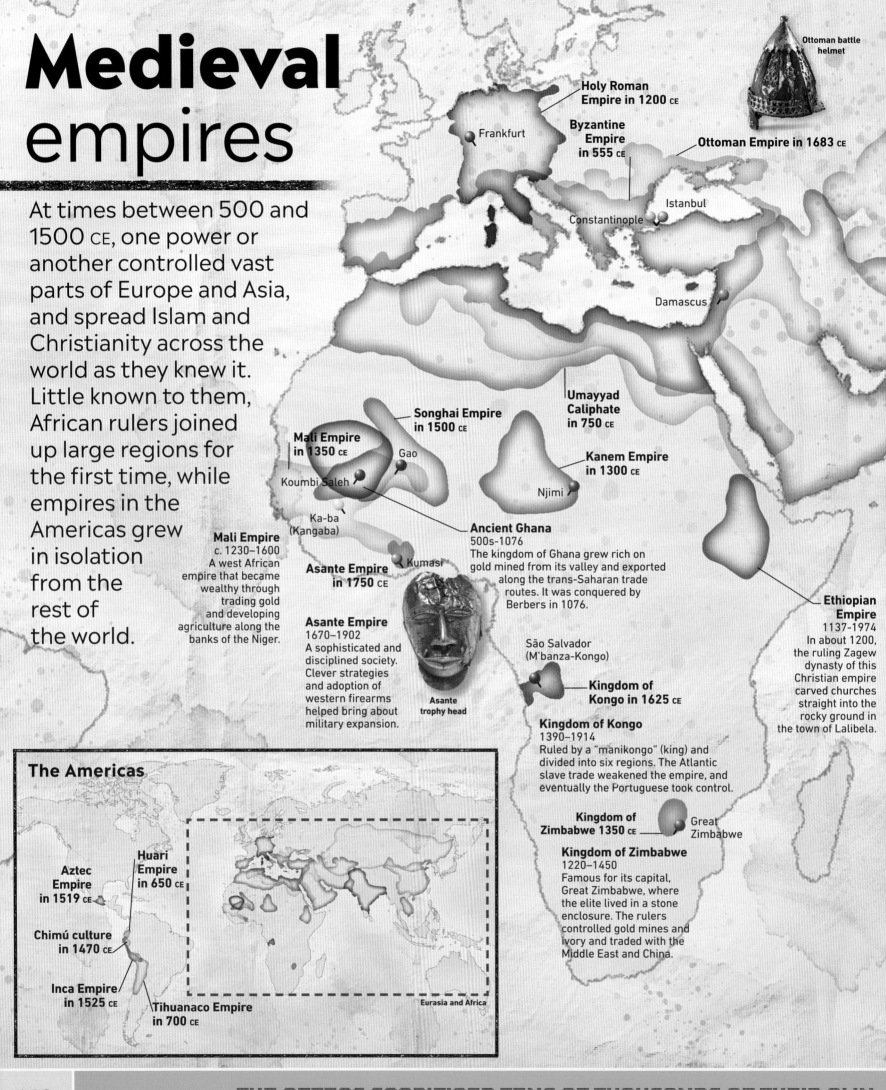

Ottoman battle helmet

Holy Roman Empire in 1200 CE

Frankfurt

Byzantine Empire in 555 CE

Ottoman Empire in 1683 CE

Istanbul

Constantinople

Damascus

Umayyad Caliphate in 750 CE

Songhai Empire in 1500 CE

Mali Empire in 1350 CE

Gao

Koumbi Saleh

Kanem Empire in 1300 CE

Njimi

Ka-ba (Kangaba)

Mali Empire
c. 1230–1600
A west African empire that became wealthy through trading gold and developing agriculture along the banks of the Niger.

Ancient Ghana
500s-1076
The kingdom of Ghana grew rich on gold mined from its valley and exported along the trans-Saharan trade routes. It was conquered by Berbers in 1076.

Kumasi

Asante Empire in 1750 CE

Asante Empire
1670–1902
A sophisticated and disciplined society. Clever strategies and adoption of western firearms helped bring about military expansion.

Asante trophy head

São Salvador (M'banza-Kongo)

Kingdom of Kongo in 1625 CE

Kingdom of Kongo
1390–1914
Ruled by a "manikongo" (king) and divided into six regions. The Atlantic slave trade weakened the empire, and eventually the Portuguese took control.

Ethiopian Empire
1137-1974
In about 1200, the ruling Zagew dynasty of this Christian empire carved churches straight into the rocky ground in the town of Lalibela.

Kingdom of Zimbabwe 1350 CE

Great Zimbabwe

Kingdom of Zimbabwe
1220–1450
Famous for its capital, Great Zimbabwe, where the elite lived in a stone enclosure. The rulers controlled gold mines and ivory and traded with the Middle East and China.

The Americas

Aztec Empire in 1519 CE

Huari Empire in 650 CE

Chimú culture in 1470 CE

Inca Empire in 1525 CE

Tihuanaco Empire in 700 CE

Eurasia and Africa

THE AZTECS SACRIFICED TENS OF THOUSANDS OF THEIR OWN

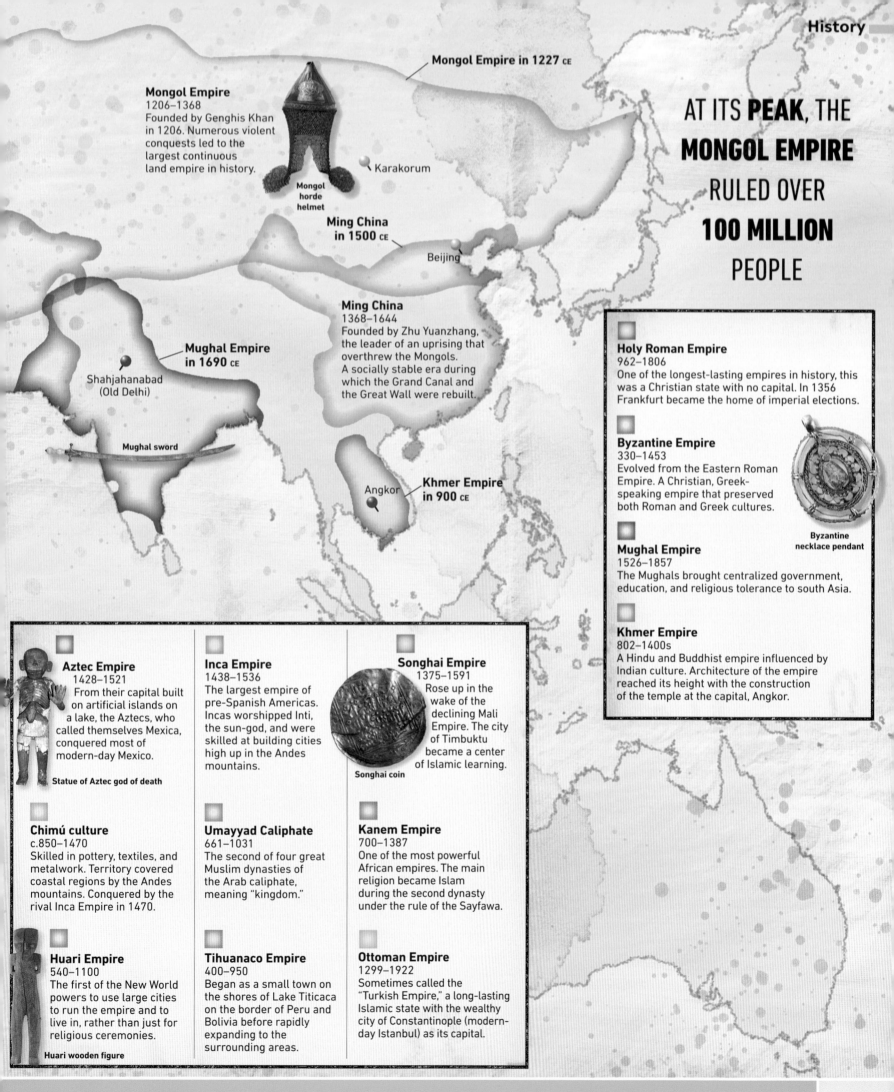

Mongol Empire in 1227 CE

Mongol Empire
1206–1368
Founded by Genghis Khan in 1206. Numerous violent conquests led to the largest continuous land empire in history.

Mongol horde helmet

Karakorum

Ming China in 1500 CE

Beijing

Ming China
1368–1644
Founded by Zhu Yuanzhang, the leader of an uprising that overthrew the Mongols. A socially stable era during which the Grand Canal and the Great Wall were rebuilt.

Mughal Empire in 1690 CE

Shahjahanabad (Old Delhi)

Mughal sword

Angkor

Khmer Empire in 900 CE

AT ITS **PEAK**, THE **MONGOL EMPIRE** RULED OVER **100 MILLION** PEOPLE

Holy Roman Empire
962–1806
One of the longest-lasting empires in history, this was a Christian state with no capital. In 1356 Frankfurt became the home of imperial elections.

Byzantine Empire
330–1453
Evolved from the Eastern Roman Empire. A Christian, Greek-speaking empire that preserved both Roman and Greek cultures.

Byzantine necklace pendant

Mughal Empire
1526–1857
The Mughals brought centralized government, education, and religious tolerance to south Asia.

Khmer Empire
802–1400s
A Hindu and Buddhist empire influenced by Indian culture. Architecture of the empire reached its height with the construction of the temple at the capital, Angkor.

Aztec Empire
1428–1521
From their capital built on artificial islands on a lake, the Aztecs, who called themselves Mexica, conquered most of modern-day Mexico.

Statue of Aztec god of death

Inca Empire
1438–1536
The largest empire of pre-Spanish Americas. Incas worshipped Inti, the sun-god, and were skilled at building cities high up in the Andes mountains.

Songhai Empire
1375–1591
Rose up in the wake of the declining Mali Empire. The city of Timbuktu became a center of Islamic learning.

Songhai coin

Chimú culture
c.850–1470
Skilled in pottery, textiles, and metalwork. Territory covered coastal regions by the Andes mountains. Conquered by the rival Inca Empire in 1470.

Umayyad Caliphate
661–1031
The second of four great Muslim dynasties of the Arab caliphate, meaning "kingdom."

Kanem Empire
700–1387
One of the most powerful African empires. The main religion became Islam during the second dynasty under the rule of the Sayfawa.

Huari Empire
540–1100
The first of the New World powers to use large cities to run the empire and to live in, rather than just for religious ceremonies.

Huari wooden figure

Tihuanaco Empire
400–950
Began as a small town on the shores of Lake Titicaca on the border of Peru and Bolivia before rapidly expanding to the surrounding areas.

Ottoman Empire
1299–1922
Sometimes called the "Turkish Empire," a long-lasting Islamic state with the wealthy city of Constantinople (modern-day Istanbul) as its capital.

PEOPLE AND ENEMY PRISONERS EACH YEAR TO APPEASE THE GODS.

Castles

From castles and forts to walled cities, rulers and nations throughout history have tried to build impregnable structures to keep their enemies at bay and strengthen their grip on power.

KEY
Flags pinpoint some of the world's most impressive fortifications.

Selected castles, forts, citadels, and fortified cities

Fort Columbia, Washington, US

Fort Union, New Mexico, US

Castillo San Felipe de Barajas, Colombia

Chan Chan ancient walled citadels, Peru

Fortifications of Valdivia, Chile

Castle of Santa Maria da Feira, Portugal

Castle of São Jorge, Portugal

Ribat of Monastir, Tunisia

Loropéni, Burkina Faso

Cape Coast Castle, Ghana

Elmina Castle, Ghana

European castles

Most were fortified residences of nobles or monarchs; others were purely defensive.

Burghausen, Germany
Europe's longest castle complex, consisting of a main castle and inner courtyard protected by five outer courtyards.

Krak des Chevaliers, Syria This 12th-century crusader castle has an outer wall with 13 towers separated from the inner wall and keep by a moat.

Coastal prisons
These two castles on Ghana's coast have a dark history: they served as fortified links along the slave trade route during the 16th century.

Forts after the age of castles

Forts became vital military centers. Their low, thick, angled walls were able to deflect cannonballs.

Asian castles

Castles in Asia reflect local building styles and look different than those in Europe, but they served the same purpose.

Himeji, Japan Built as a fort in 1333, Himeji was then rebuilt several times between the 14th and 17th centuries. It has 83 buildings protected by 85-ft- (26-m-) high walls and 3 moats, and is Japan's largest castle.

Mehrangarh Fort, India This fort, 400 ft (122 m) above the city of Jodhpur, hides several palaces within its walls. Built by the ruler Rao Jodha in 1459, it is entered through a series of seven gates.

Fortified cities

Cities surrounded by defensive walls, often incorporating a castle or royal residence.

Forbidden City, China The former imperial palace in Beijing has 980 buildings ringed by a wall and a 171-ft- (52-m-) wide moat.

Great Zimbabwe Once the capital of the Kingdom of Zimbabwe, the stone walls of this royal city were built without using mortar.

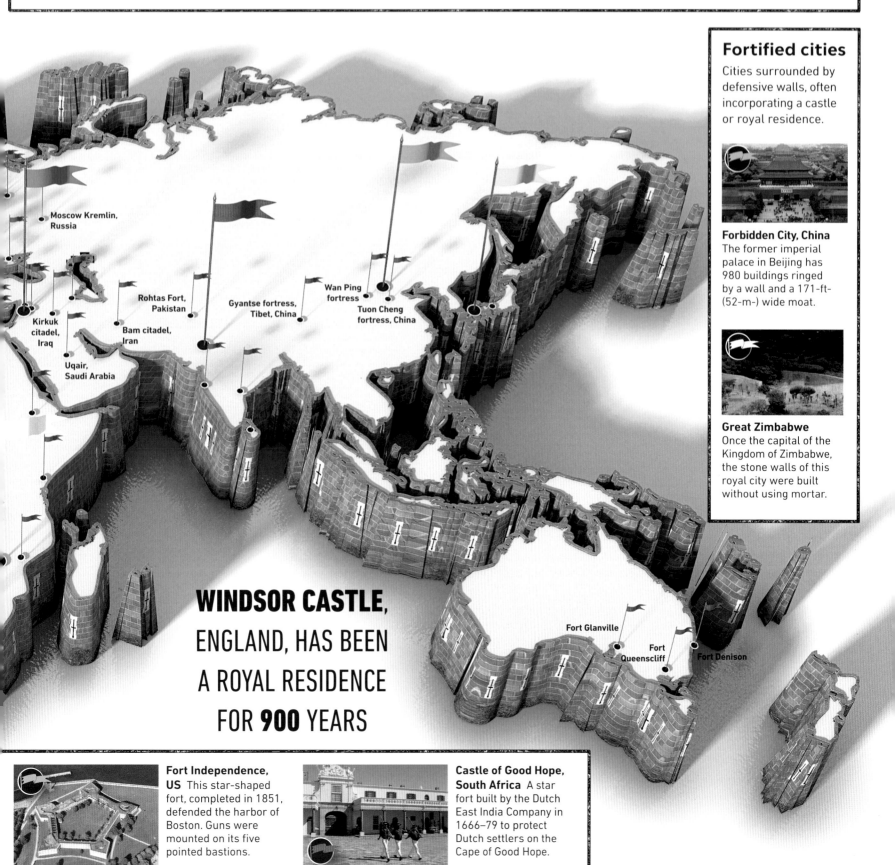

Moscow Kremlin, Russia

Kirkuk citadel, Iraq

Uqair, Saudi Arabia

Bam citadel, Iran

Rohtas Fort, Pakistan

Gyantse fortress, Tibet, China

Wan Ping fortress

Tuon Cheng fortress, China

Fort Glanville

Fort Queenscliff

Fort Denison

WINDSOR CASTLE, ENGLAND, HAS BEEN A ROYAL RESIDENCE FOR **900** YEARS

Fort Independence, US This star-shaped fort, completed in 1851, defended the harbor of Boston. Guns were mounted on its five pointed bastions.

Castle of Good Hope, South Africa A star fort built by the Dutch East India Company in 1666–79 to protect Dutch settlers on the Cape of Good Hope.

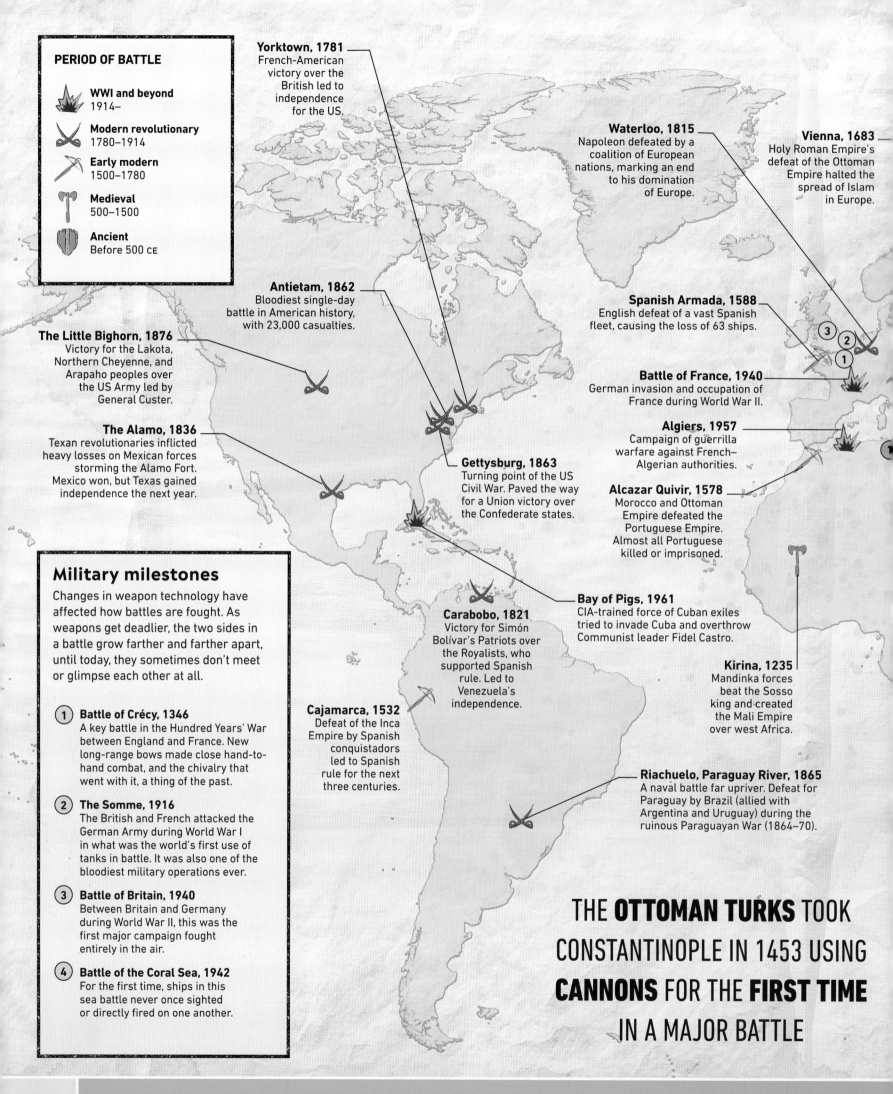

PERIOD OF BATTLE

WWI and beyond
1914–

Modern revolutionary
1780–1914

Early modern
1500–1780

Medieval
500–1500

Ancient
Before 500 CE

Yorktown, 1781
French-American victory over the British led to independence for the US.

Waterloo, 1815
Napoleon defeated by a coalition of European nations, marking an end to his domination of Europe.

Vienna, 1683
Holy Roman Empire's defeat of the Ottoman Empire halted the spread of Islam in Europe.

Antietam, 1862
Bloodiest single-day battle in American history, with 23,000 casualties.

Spanish Armada, 1588
English defeat of a vast Spanish fleet, causing the loss of 63 ships.

The Little Bighorn, 1876
Victory for the Lakota, Northern Cheyenne, and Arapaho peoples over the US Army led by General Custer.

Battle of France, 1940
German invasion and occupation of France during World War II.

Algiers, 1957
Campaign of guerrilla warfare against French–Algerian authorities.

The Alamo, 1836
Texan revolutionaries inflicted heavy losses on Mexican forces storming the Alamo Fort. Mexico won, but Texas gained independence the next year.

Gettysburg, 1863
Turning point of the US Civil War. Paved the way for a Union victory over the Confederate states.

Alcazar Quivir, 1578
Morocco and Ottoman Empire defeated the Portuguese Empire. Almost all Portuguese killed or imprisoned.

Military milestones

Changes in weapon technology have affected how battles are fought. As weapons get deadlier, the two sides in a battle grow farther and farther apart, until today, they sometimes don't meet or glimpse each other at all.

Bay of Pigs, 1961
CIA-trained force of Cuban exiles tried to invade Cuba and overthrow Communist leader Fidel Castro.

Carabobo, 1821
Victory for Simón Bolívar's Patriots over the Royalists, who supported Spanish rule. Led to Venezuela's independence.

Kirina, 1235
Mandinka forces beat the Sosso king and created the Mali Empire over west Africa.

① Battle of Crécy, 1346
A key battle in the Hundred Years' War between England and France. New long-range bows made close hand-to-hand combat, and the chivalry that went with it, a thing of the past.

Cajamarca, 1532
Defeat of the Inca Empire by Spanish conquistadors led to Spanish rule for the next three centuries.

② The Somme, 1916
The British and French attacked the German Army during World War I in what was the world's first use of tanks in battle. It was also one of the bloodiest military operations ever.

Riachuelo, Paraguay River, 1865
A naval battle far upriver. Defeat for Paraguay by Brazil (allied with Argentina and Uruguay) during the ruinous Paraguayan War (1864–70).

③ Battle of Britain, 1940
Between Britain and Germany during World War II, this was the first major campaign fought entirely in the air.

④ Battle of the Coral Sea, 1942
For the first time, ships in this sea battle never once sighted or directly fired on one another.

THE **OTTOMAN TURKS** TOOK CONSTANTINOPLE IN 1453 USING **CANNONS** FOR THE **FIRST TIME** IN A MAJOR BATTLE

Sieges

Not strictly a battle, a siege is a military blockade of a city or fortress. The aim is to conquer the city by waiting for those inside to surrender. Sometimes, the side laying siege attacks to speed things up.

(1) Siege of Carthage 149–146 BCE
One of the longest sieges in history. The Romans surrounded Carthage (in modern Tunisia) and waited 3 years for its surrender, then enslaved the Carthaginian population.

(2) Capture of Jerusalem, 1099
During the Crusader wars between Christians and Muslims, the Muslim defenders of Jerusalem lost control when the Christians built two enormous siege engines (towers on wheels) and scaled the walls.

Austerlitz, 1805
With smaller forces, the French Empire crushed Russia and Austria. One of Napoleon's greatest victories.

Actium, 31 BCE
Rome declared war on Antony and Cleopatra of Egypt. The Roman victory led to the beginning of the Roman Empire.

Thermopylae, 480 BCE
Vastly outnumbered Greek forces held the Persian Emperor Xerxes at bay for a vital 3 days.

Stalingrad, 1942–43
Long siege of this Soviet city caused immense suffering on both sides and eventually led to crippling defeat for Nazi Germany.

Fall of Constantinople, 1453
After a 4-month siege, Byzantine Empire fell to the invading Ottoman Empire.

Badger Mouth, 1211
Mongol ruler Genghis Khan's victory over the Jin Dynasty of China. One of history's bloodiest battles.

Huai-Hai, 1948
Final major fight in Chinese Civil War that led to the Communist takeover of China.

Battle of Inchon, 1950
A clear victory for the United Nations against North Korean forces in the Korean War.

Iwo Jima, 1945
The US captured this island as a way of possibly invading Japan. More than 21,000 Japanese died.

Wuhan, 1938
Soviet and revolutionary Chinese forces totaling 1,100,000 troops and 200 aircraft failed to stop Japan from capturing the city.

Battle of Phillora, 1965
One of the largest tank battles of the Indo-Pakistani War. Decisive victory for Indian Army.

Omdurman, 1898
Small British and Egyptian forces massacred a huge, but ill-equipped, Sudanese Army.

El Alamein, 1942
Major tank battle of World War II. British-led victory over Axis Powers (Italy and Germany).

Dien Bien Phu, 1954
Viet Minh communist revolutionaries besieged and defeated the French to end the First Indochina War. The next year began another 20 years of fighting in Vietnam.

Kalinga, 262–261 BCE
The Mauryan Empire under Ashoka the Great fought the republic of Kalinga. At least 100,000 Kalingans were killed.

Isandlwana, 1879
Crushing victory for the Zulu nation over the British, despite relying mainly on spears and cowhide shields.

Surabaya, 1945
Heaviest battle of the Indonesian Revolution against the British and Dutch. Celebrated as Heroes' Day in Indonesia.

Coral Sea, 1942
World War II naval battle between Japan and the US and Australia. The battle was the first time aircraft carriers engaged each other.

Battlegrounds

At one time, armies met in formation on a single field of battle and fought for one to several days. By the 20th century, long-range weapons had changed warfare. Battlefields in places became theaters of war the size of countries.

IS SAID TO HAVE REJECTED WARFARE AND TURNED TO BUDDHISM.

The last empires

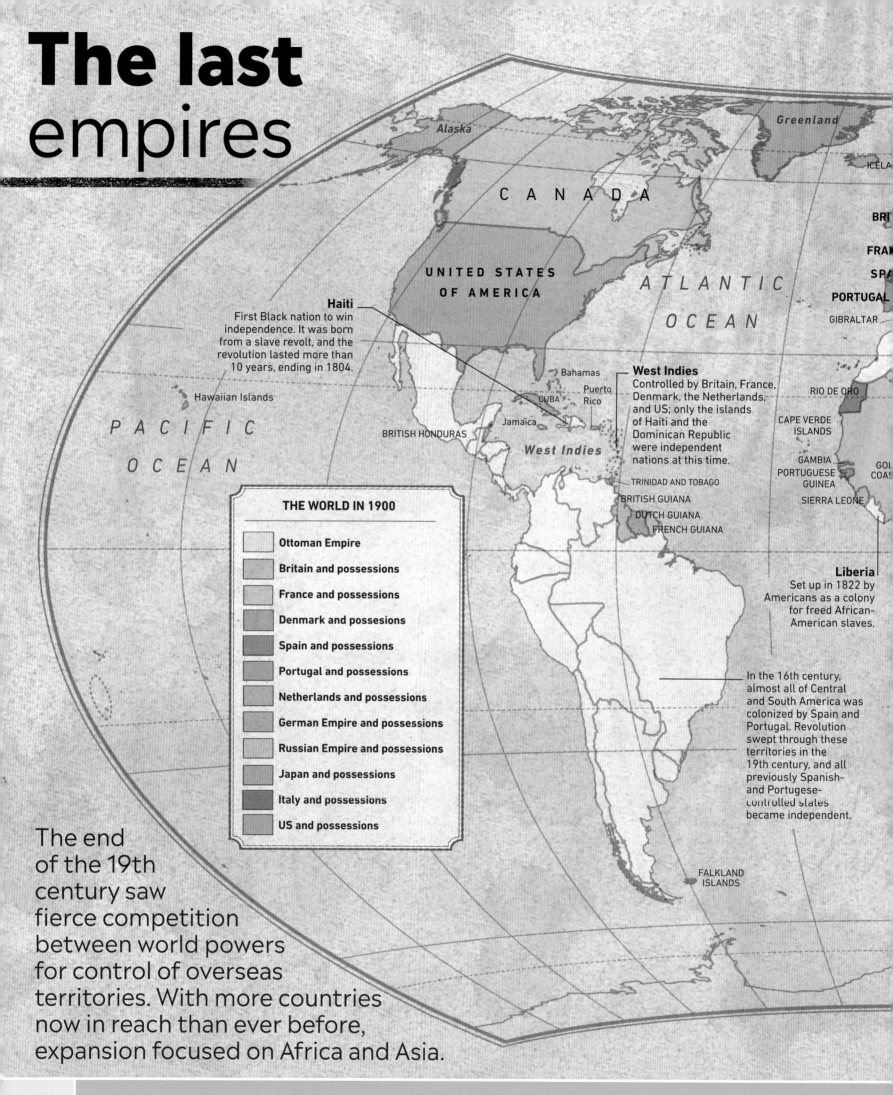

Alaska

Greenland

ICELA

C A N A D A

BRI

UNITED STATES
OF AMERICA

A T L A N T I C

FRA

SPA

PORTUGAL

O C E A N

GIBRALTAR

Haiti
First Black nation to win
independence. It was born
from a slave revolt, and the
revolution lasted more than
10 years, ending in 1804.

Bahamas

Puerto
Rico

CUBA

RIO DE ORO

Hawaiian Islands

West Indies
Controlled by Britain, France,
Denmark, the Netherlands,
and US; only the islands
of Haiti and the
Dominican Republic
were independent
nations at this time.

P A C I F I C

Jamaica

CAPE VERDE
ISLANDS

BRITISH HONDURAS

O C E A N

West Indies

GAMBIA
PORTUGUESE
GUINEA

GOL
COAS

TRINIDAD AND TOBAGO

SIERRA LEONE

BRITISH GUIANA

DUTCH GUIANA

FRENCH GUIANA

THE WORLD IN 1900

- Ottoman Empire
- Britain and possessions
- France and possessions
- Denmark and possesions
- Spain and possessions
- Portugal and possessions
- Netherlands and possessions
- German Empire and possessions
- Russian Empire and possessions
- Japan and possessions
- Italy and possessions
- US and possessions

Liberia
Set up in 1822 by
Americans as a colony
for freed African-
American slaves.

In the 16th century,
almost all of Central
and South America was
colonized by Spain and
Portugal. Revolution
swept through these
territories in the
19th century, and all
previously Spanish-
and Portugese-
controlled states
became independent.

FALKLAND
ISLANDS

The end
of the 19th
century saw
fierce competition
between world powers
for control of overseas
territories. With more countries
now in reach than ever before,
expansion focused on Africa and Asia.

AT ITS HEIGHT IN 1922, THE BRITISH EMPIRE CONTROLLED

IN **1900, EUROPEANS** CONTROLLED **90 PERCENT** OF **AFRICA**

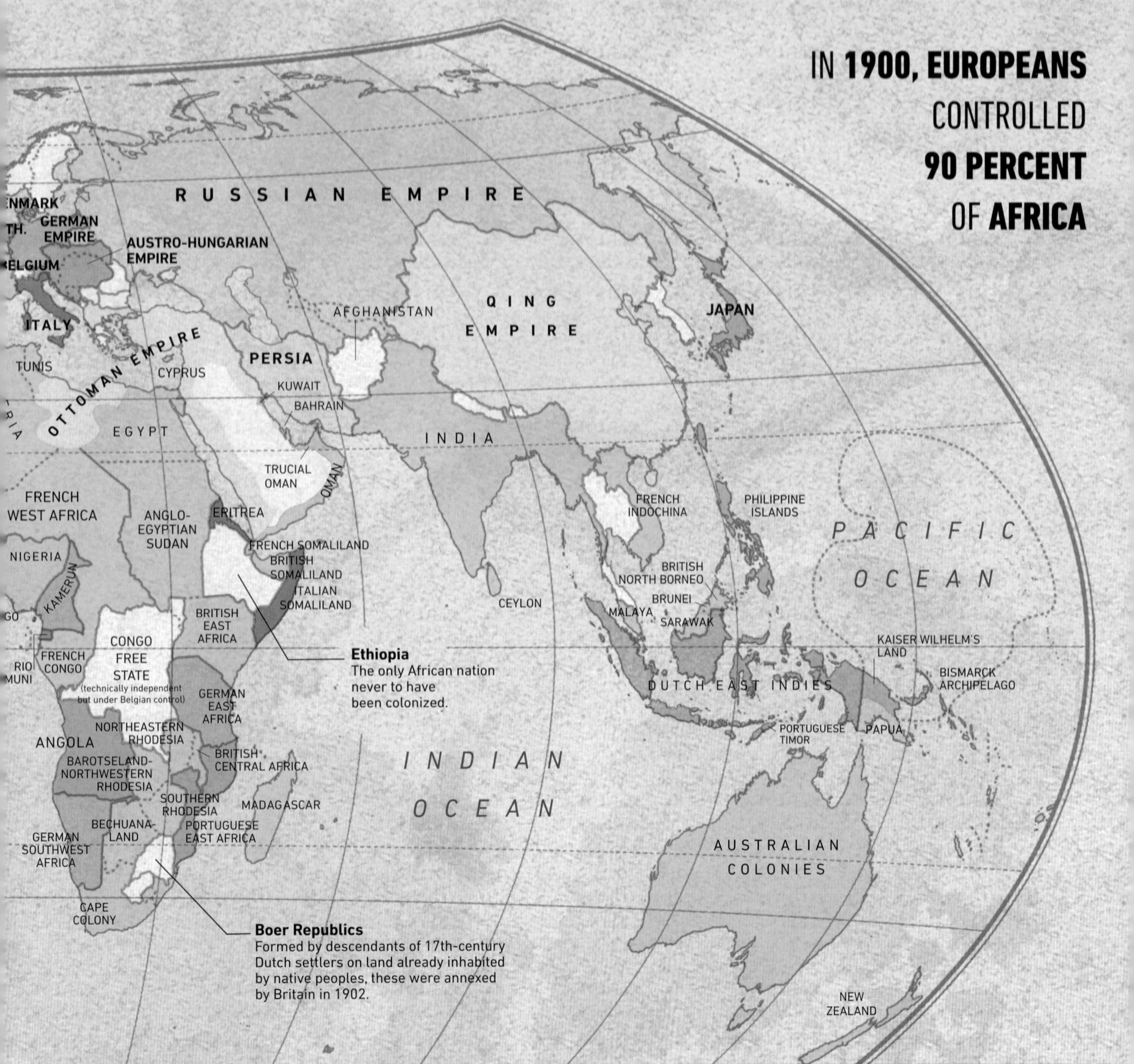
RUSSIAN EMPIRE
GERMAN EMPIRE
AUSTRO-HUNGARIAN EMPIRE
ITALY
TUNIS
OTTOMAN EMPIRE
CYPRUS
PERSIA
KUWAIT
BAHRAIN
AFGHANISTAN
QING EMPIRE
JAPAN
EGYPT
INDIA
TRUCIAL OMAN
OMAN
FRENCH WEST AFRICA
ANGLO-EGYPTIAN SUDAN
ERITREA
FRENCH SOMALILAND
BRITISH SOMALILAND
ITALIAN SOMALILAND
NIGERIA
KAMERUN
BRITISH EAST AFRICA
CONGO FREE STATE (technically independent but under Belgian control)
FRENCH CONGO
RIO MUNI
GERMAN EAST AFRICA
Ethiopia
The only African nation never to have been colonized.
NORTHEASTERN RHODESIA
ANGOLA
BAROTSELAND-NORTHWESTERN RHODESIA
BRITISH CENTRAL AFRICA
SOUTHERN RHODESIA
MADAGASCAR
BECHUANA-LAND
PORTUGUESE EAST AFRICA
GERMAN SOUTHWEST AFRICA
CAPE COLONY
Boer Republics
Formed by descendants of 17th-century Dutch settlers on land already inhabited by native peoples, these were annexed by Britain in 1902.
CEYLON
FRENCH INDOCHINA
PHILIPPINE ISLANDS
BRITISH NORTH BORNEO
BRUNEI
MALAYA
SARAWAK
PACIFIC OCEAN
KAISER WILHELM'S LAND
BISMARCK ARCHIPELAGO
DUTCH EAST INDIES
PORTUGUESE TIMOR
PAPUA
INDIAN OCEAN
AUSTRALIAN COLONIES
NEW ZEALAND

Scramble for Africa

The Atlantic slave trade, in which Africans were forcibly sold to people in the Americas, ended in the mid-19th century. European powers colonized Africa for economic, political, and religious reasons, scrambling to claim territory before their rivals.

- **1871:** Germany and Italy are both unified. No more territory available for expansion of empires in Europe.
- **1884–85:** Berlin Conference, where European powers decide rules on carving up Africa.
- **1900:** Only a handful of regions are still independent states. Britain rules 30 percent of Africa's population.

The Great Game

In the 1830s, Britain feared Russia was planning on invading British-ruled India through controlling India's neighbor, Afghanistan. The "Great Game" was the rivalry for power in Asia between the British and Russian empires.

- **1839–42:** First Anglo-Afghan War. Terrible defeat at Kabul for the British.
- **1878–80:** Second Anglo-Afghan War. Russia is defeated and Britain withdraws but takes control of Afghanistan's foreign affairs.
- **1907:** Russia and Britain sign a peace treaty in the face of the German threat of expansion in the Middle East.

ALMOST ONE-QUARTER OF THE WORLD'S POPULATION AND LAND.

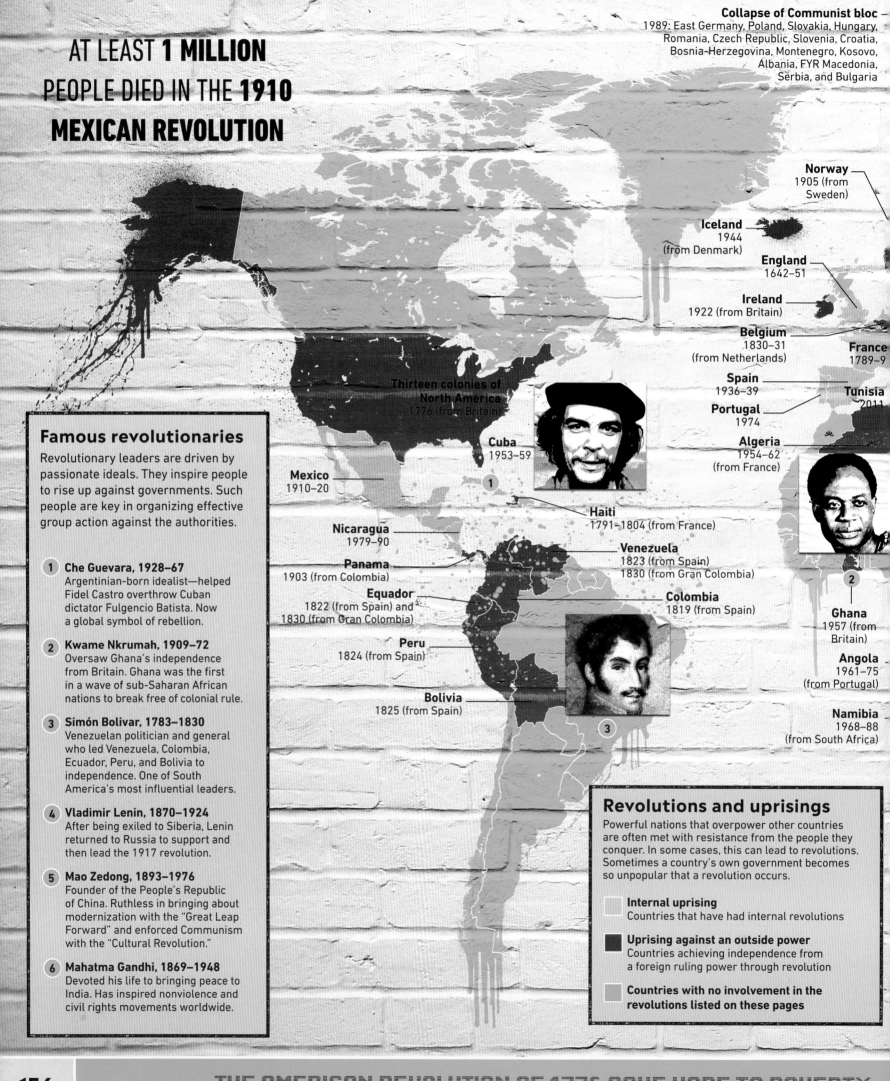

AT LEAST 1 MILLION PEOPLE DIED IN THE 1910 MEXICAN REVOLUTION

Collapse of Communist bloc
1989: East Germany, Poland, Slovakia, Hungary, Romania, Czech Republic, Slovenia, Croatia, Bosnia-Herzegovina, Montenegro, Kosovo, Albania, FYR Macedonia, Serbia, and Bulgaria

Norway
1905 (from Sweden)

Iceland
1944
(from Denmark)

England
1642–51

Ireland
1922 (from Britain)

Belgium
1830–31
(from Netherlands)

France
1789–9

Spain
1936–39

Portugal
1974

Tunisia
2011

Algeria
1954–62
(from France)

Thirteen colonies of North America
1776 (from Britain)

Cuba
1953–59

Famous revolutionaries

Revolutionary leaders are driven by passionate ideals. They inspire people to rise up against governments. Such people are key in organizing effective group action against the authorities.

1 Che Guevara, 1928–67
Argentinian-born idealist—helped Fidel Castro overthrow Cuban dictator Fulgencio Batista. Now a global symbol of rebellion.

2 Kwame Nkrumah, 1909–72
Oversaw Ghana's independence from Britain. Ghana was the first in a wave of sub-Saharan African nations to break free of colonial rule.

3 Simón Bolivar, 1783–1830
Venezuelan politician and general who led Venezuela, Colombia, Ecuador, Peru, and Bolivia to independence. One of South America's most influential leaders.

4 Vladimir Lenin, 1870–1924
After being exiled to Siberia, Lenin returned to Russia to support and then lead the 1917 revolution.

5 Mao Zedong, 1893–1976
Founder of the People's Republic of China. Ruthless in bringing about modernization with the "Great Leap Forward" and enforced Communism with the "Cultural Revolution."

6 Mahatma Gandhi, 1869–1948
Devoted his life to bringing peace to India. Has inspired nonviolence and civil rights movements worldwide.

Mexico
1910–20

Nicaragua
1979–90

Panama
1903 (from Colombia)

Equador
1822 (from Spain) and 1830 (from Gran Colombia)

Peru
1824 (from Spain)

Bolivia
1825 (from Spain)

Haiti
1791–1804 (from France)

Venezuela
1823 (from Spain)
1830 (from Gran Colombia)

Colombia
1819 (from Spain)

Ghana
1957 (from Britain)

Angola
1961–75
(from Portugal)

Namibia
1968–88
(from South Africa)

Revolutions and uprisings

Powerful nations that overpower other countries are often met with resistance from the people they conquer. In some cases, this can lead to revolutions. Sometimes a country's own government becomes so unpopular that a revolution occurs.

Internal uprising
Countries that have had internal revolutions

Uprising against an outside power
Countries achieving independence from a foreign ruling power through revolution

Countries with no involvement in the revolutions listed on these pages

Revolutions

People all over the world have risen up against oppressive rulers. Revolutions can be sudden or lengthy, bloody or peaceful, but have one thing in common: they are all an attempt to change the way a country is ruled.

Lithuania
1989 (from USSR)

Finland
1917 (from Russia)

Estonia
1989 (from USSR)

Latvia
1989 (from USSR)

Belarus
1989 (from USSR)

Ukraine
1989 (from USSR)

Moldova
1989 (from USSR)

Georgia
1989 (from USSR)

Armenia
1989 (from USSR)

Greece
1821–32 (from Ottoman Empire)

Turkmenistan
1989 (from USSR)

Uzbekistan
1989 (from USSR)

Kazakhstan
1989 (from USSR)

Kyrgyzstan
1989 (from USSR)

Tajikistan
1989 (from USSR)

Russia
1917

4

5

Korea
1945
(from Japan)

Iraq
2014–17

Iran
1979

Afghanistan
1996

Azerbaijan
1989 (from USSR)

Syria
From 2011

Libya
2011

Egypt
2011

Eritrea
1961–91
(from Ethiopia)

South Sudan
2011 (from Sudan)

Democratic Republic of the Congo
1997

Rwanda
1961
(from Belgium)

Yemen
2011

India
1947 (from Britain)

6

Somalia
1986–92

Kenya (Mau Mau)
1952–60 (from Britain)

Madagascar
1960 (from France)

China
1949

Myanmar
(Burma)
1962

Laos
1975
(from USSR)

Vietnam
1975 (Socialist Republic of Vietnam created after war between North and South Vietnam)

Philippines
1896–98 (from Spain)

Cambodia
1979 (Khmer Rouge)

Singapore
1965 (from Malaysia)

Papua New Guinea
1975 (from Australia)

Indonesia
1945–49
(from the Netherlands)

East Timor
1975 (from Portugal) and 2002
(from Indonesia)

South Africa
1994

Collapse of Communism

The USSR was a Communist state that incorporated Russia and 14 other Soviet republics (some of the red areas on the map). The USSR also had great influence over several other European states that collectively were known as the "Communist bloc" (some of the yellow map areas). In 1989, revolution spread through all these states, and in 1991 the USSR was dissolved.

▼ **Fall of communism**
Indicates countries in which Communism collapsed in 1989–91

Arab Spring

The "Arab Spring" revolutions and protests swept through the Arab world in 2011. As the map shows, in some countries rulers were forced out, while in others there were failed uprisings. The Arab Spring was the first uprising where protestors used social media to coordinate their actions. Not all of the movements were successful, however; the uprising in Tunisia led to a number of improvements, but many of the other countries are still marked by unrest.

Arab Spring
Indicates countries involved in the Arab Spring

STRICKEN EUROPE AND INSPIRED THE 1789-99 FRENCH REVOLUTION.

Scapa Flow (1919)
After World War I, the German navy sank 52 of its own ships here, rather than surrender them to Britain.

SS *Islander* (1901)
Its cargo of gold, which some estimate is worth up to $700 million today, has never been found.

SS *Sultana* (1865)
This river steamer exploded in the Mississippi River with the loss of about 1,700 lives.

Méduse (1816)
When the *Méduse* sank, 147 crewmen built a life raft, but only 15 survived to be rescued.

HMS *Agamemnon* (1809)
A former command of Admiral Nelson, she struck an uncharted group of rocks in a bay off Uruguay.

Shipwrecks

The beds and shores of the world's seas, lakes, and rivers are littered with shipwrecks. Some are famous either for the huge loss of life they caused or the enormous value of their cargo.

Natural shipwrecks

Sailors battle constantly against the phenomenal forces of nature, and one of the most common causes of shipwrecks is bad weather. Storms and hurricanes batter ships and blow them off course, and fog, rain, or snow reduce visibility. Ice is another big risk. An iceberg can inflict fatal damage to a ship if it collides with one; while ice that builds up on the body of a ship can also cause it to become unstable and capsize.

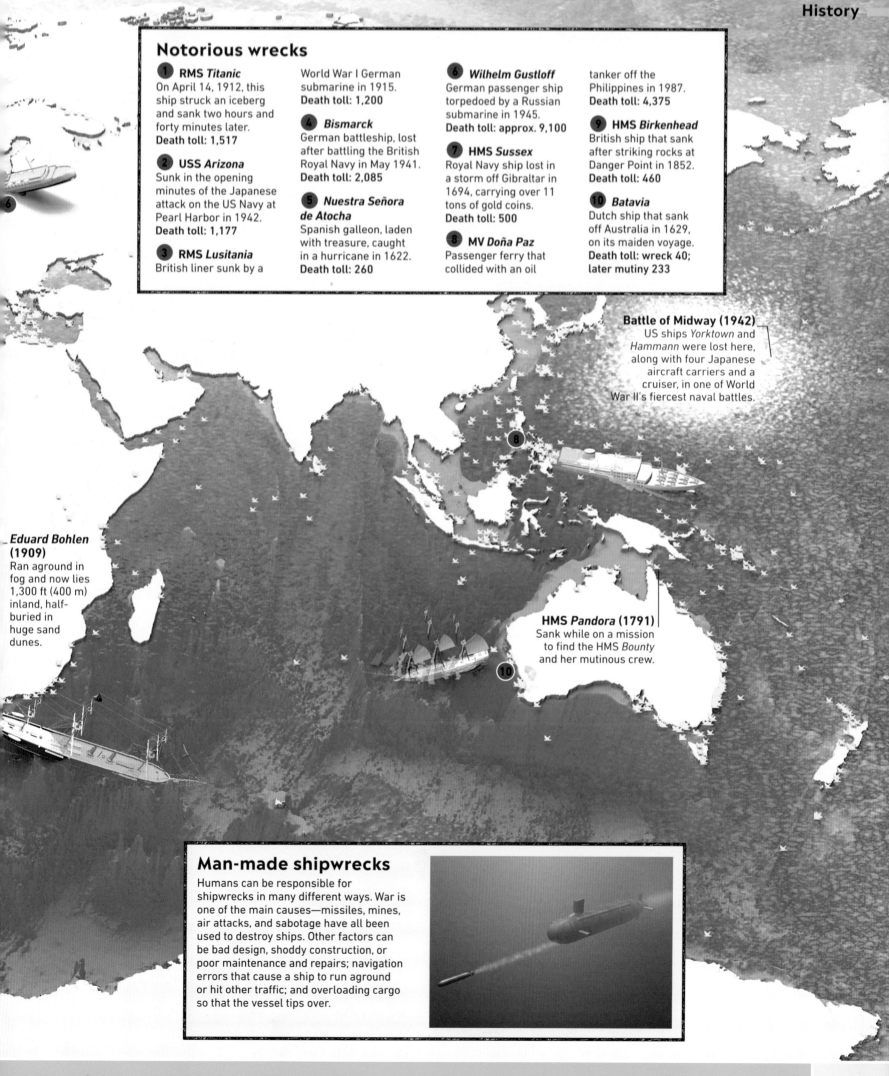

Notorious wrecks

1 RMS Titanic
On April 14, 1912, this ship struck an iceberg and sank two hours and forty minutes later.
Death toll: 1,517

2 USS Arizona
Sunk in the opening minutes of the Japanese attack on the US Navy at Pearl Harbor in 1942.
Death toll: 1,177

3 RMS Lusitania
British liner sunk by a World War I German submarine in 1915.
Death toll: 1,200

4 Bismarck
German battleship, lost after battling the British Royal Navy in May 1941.
Death toll: 2,085

5 Nuestra Señora de Atocha
Spanish galleon, laden with treasure, caught in a hurricane in 1622.
Death toll: 260

6 Wilhelm Gustloff
German passenger ship torpedoed by a Russian submarine in 1945.
Death toll: approx. 9,100

7 HMS Sussex
Royal Navy ship lost in a storm off Gibraltar in 1694, carrying over 11 tons of gold coins.
Death toll: 500

8 MV Doña Paz
Passenger ferry that collided with an oil tanker off the Philippines in 1987.
Death toll: 4,375

9 HMS Birkenhead
British ship that sank after striking rocks at Danger Point in 1852.
Death toll: 460

10 Batavia
Dutch ship that sank off Australia in 1629, on its maiden voyage.
Death toll: wreck 40; later mutiny 233

Battle of Midway (1942)
US ships *Yorktown* and *Hammann* were lost here, along with four Japanese aircraft carriers and a cruiser, in one of World War II's fiercest naval battles.

Eduard Bohlen (1909)
Ran aground in fog and now lies 1,300 ft (400 m) inland, half-buried in huge sand dunes.

HMS Pandora (1791)
Sank while on a mission to find the HMS *Bounty* and her mutinous crew.

Man-made shipwrecks

Humans can be responsible for shipwrecks in many different ways. War is one of the main causes—missiles, mines, air attacks, and sabotage have all been used to destroy ships. Other factors can be bad design, shoddy construction, or poor maintenance and repairs; navigation errors that cause a ship to run aground or hit other traffic; and overloading cargo so that the vessel tips over.

SANK IN 1978, SPILLING MORE THAN 220,000 TONS OF CRUDE OIL.

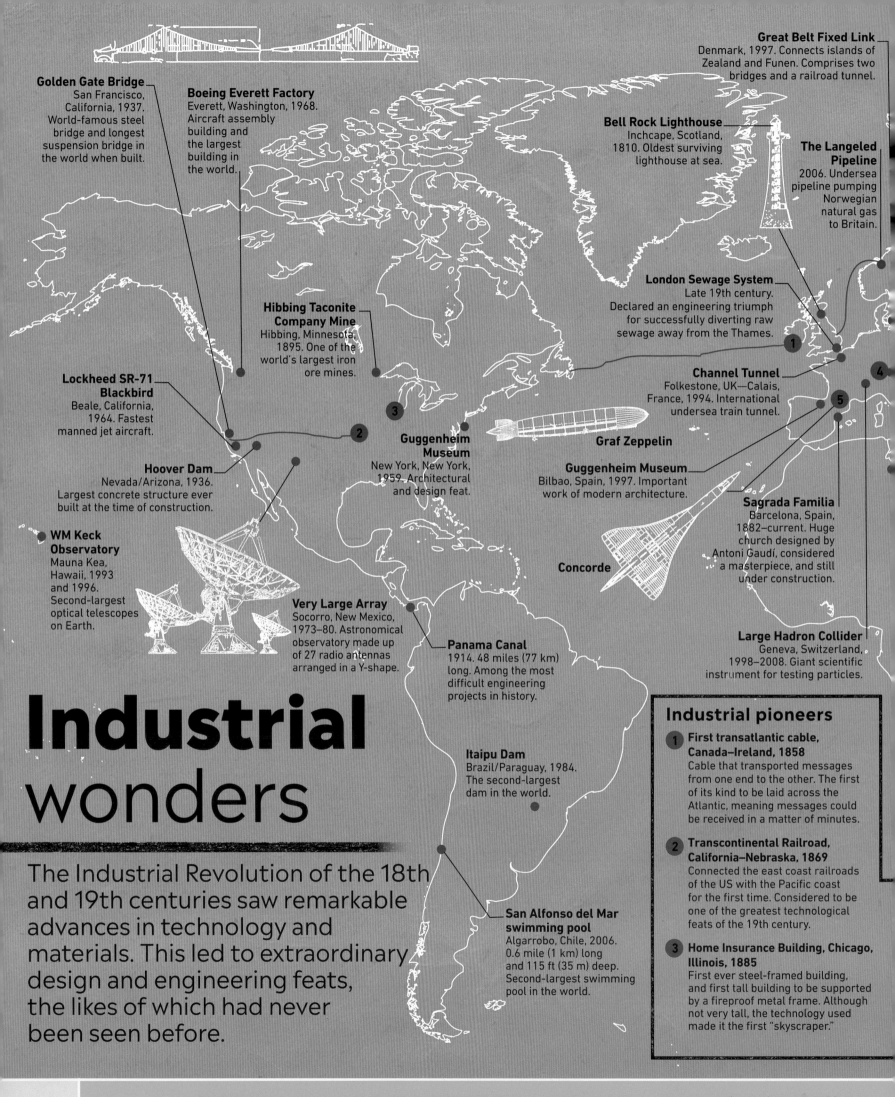

Industrial wonders

The Industrial Revolution of the 18th and 19th centuries saw remarkable advances in technology and materials. This led to extraordinary design and engineering feats, the likes of which had never been seen before.

Golden Gate Bridge
San Francisco, California, 1937. World-famous steel bridge and longest suspension bridge in the world when built.

Boeing Everett Factory
Everett, Washington, 1968. Aircraft assembly building and the largest building in the world.

Great Belt Fixed Link
Denmark, 1997. Connects islands of Zealand and Funen. Comprises two bridges and a railroad tunnel.

Bell Rock Lighthouse
Inchcape, Scotland, 1810. Oldest surviving lighthouse at sea.

The Langeled Pipeline
2006. Undersea pipeline pumping Norwegian natural gas to Britain.

London Sewage System
Late 19th century. Declared an engineering triumph for successfully diverting raw sewage away from the Thames.

Hibbing Taconite Company Mine
Hibbing, Minnesota, 1895. One of the world's largest iron ore mines.

Lockheed SR-71 Blackbird
Beale, California, 1964. Fastest manned jet aircraft.

Channel Tunnel
Folkestone, UK—Calais, France, 1994. International undersea train tunnel.

Hoover Dam
Nevada/Arizona, 1936. Largest concrete structure ever built at the time of construction.

Graf Zeppelin

Guggenheim Museum
New York, New York, 1959. Architectural and design feat.

Guggenheim Museum
Bilbao, Spain, 1997. Important work of modern architecture.

WM Keck Observatory
Mauna Kea, Hawaii, 1993 and 1996. Second-largest optical telescopes on Earth.

Sagrada Familia
Barcelona, Spain, 1882–current. Huge church designed by Antoni Gaudí, considered a masterpiece, and still under construction.

Concorde

Very Large Array
Socorro, New Mexico, 1973–80. Astronomical observatory made up of 27 radio antennas arranged in a Y-shape.

Panama Canal
1914. 48 miles (77 km) long. Among the most difficult engineering projects in history.

Large Hadron Collider
Geneva, Switzerland, 1998–2008. Giant scientific instrument for testing particles.

Itaipu Dam
Brazil/Paraguay, 1984. The second-largest dam in the world.

San Alfonso del Mar swimming pool
Algarrobo, Chile, 2006. 0.6 mile (1 km) long and 115 ft (35 m) deep. Second-largest swimming pool in the world.

Industrial pioneers

(1) First transatlantic cable, Canada–Ireland, 1858
Cable that transported messages from one end to the other. The first of its kind to be laid across the Atlantic, meaning messages could be received in a matter of minutes.

(2) Transcontinental Railroad, California–Nebraska, 1869
Connected the east coast railroads of the US with the Pacific coast for the first time. Considered to be one of the greatest technological feats of the 19th century.

(3) Home Insurance Building, Chicago, Illinois, 1885
First ever steel-framed building, and first tall building to be supported by a fireproof metal frame. Although not very tall, the technology used made it the first "skyscraper."

FRENCH ENGINEER ALBERT MATHIEU PUT FORWARD A PROPOSAL

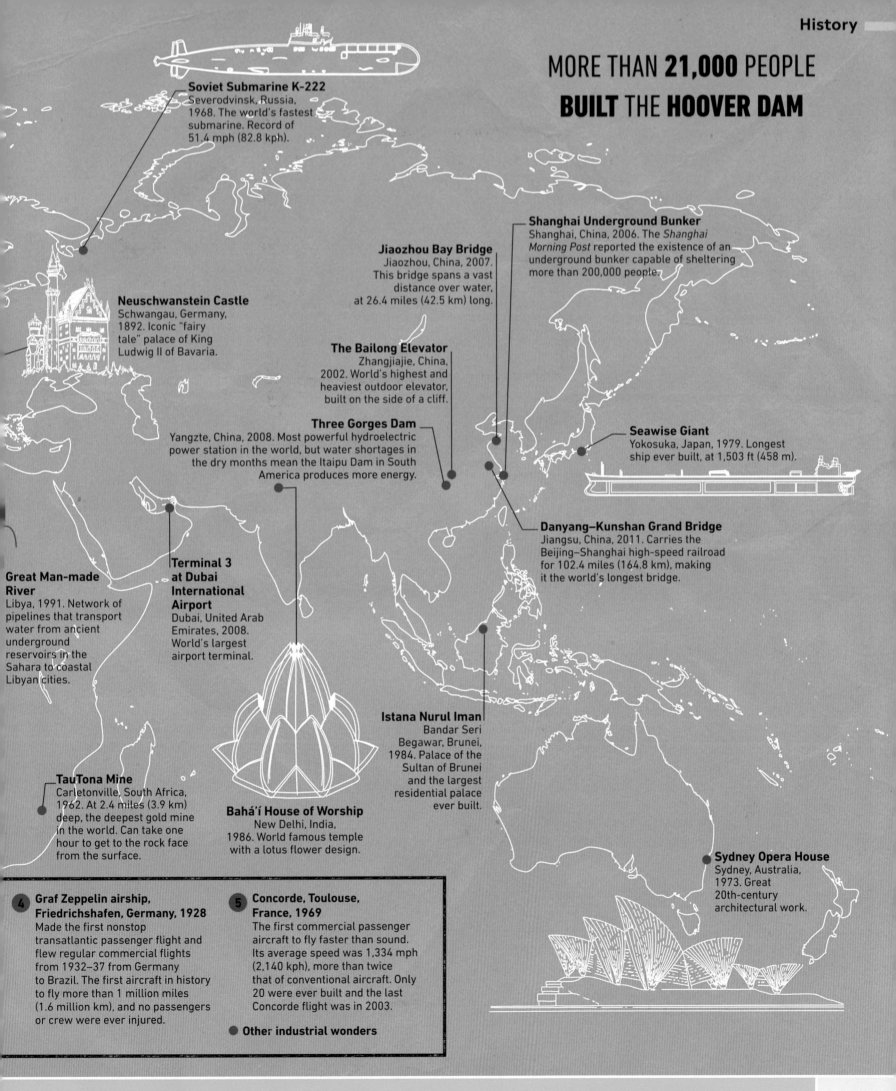

MORE THAN **21,000** PEOPLE **BUILT** THE **HOOVER DAM**

Soviet Submarine K-222
Severodvinsk, Russia, 1968. The world's fastest submarine. Record of 51.4 mph (82.8 kph).

Shanghai Underground Bunker
Shanghai, China, 2006. The *Shanghai Morning Post* reported the existence of an underground bunker capable of sheltering more than 200,000 people.

Jiaozhou Bay Bridge
Jiaozhou, China, 2007. This bridge spans a vast distance over water, at 26.4 miles (42.5 km) long.

Neuschwanstein Castle
Schwangau, Germany, 1892. Iconic "fairy tale" palace of King Ludwig II of Bavaria.

The Bailong Elevator
Zhangjiajie, China, 2002. World's highest and heaviest outdoor elevator, built on the side of a cliff.

Three Gorges Dam
Yangzte, China, 2008. Most powerful hydroelectric power station in the world, but water shortages in the dry months mean the Itaipu Dam in South America produces more energy.

Seawise Giant
Yokosuka, Japan, 1979. Longest ship ever built, at 1,503 ft (458 m).

Great Man-made River
Libya, 1991. Network of pipelines that transport water from ancient underground reservoirs in the Sahara to coastal Libyan cities.

Terminal 3 at Dubai International Airport
Dubai, United Arab Emirates, 2008. World's largest airport terminal.

Danyang–Kunshan Grand Bridge
Jiangsu, China, 2011. Carries the Beijing–Shanghai high-speed railroad for 102.4 miles (164.8 km), making it the world's longest bridge.

Istana Nurul Iman
Bandar Seri Begawar, Brunei, 1984. Palace of the Sultan of Brunei and the largest residential palace ever built.

TauTona Mine
Carletonville, South Africa, 1962. At 2.4 miles (3.9 km) deep, the deepest gold mine in the world. Can take one hour to get to the rock face from the surface.

Bahá'í House of Worship
New Delhi, India, 1986. World famous temple with a lotus flower design.

Sydney Opera House
Sydney, Australia, 1973. Great 20th-century architectural work.

4 **Graf Zeppelin airship, Friedrichshafen, Germany, 1928**
Made the first nonstop transatlantic passenger flight and flew regular commercial flights from 1932–37 from Germany to Brazil. The first aircraft in history to fly more than 1 million miles (1.6 million km), and no passengers or crew were ever injured.

5 **Concorde, Toulouse, France, 1969**
The first commercial passenger aircraft to fly faster than sound. Its average speed was 1,334 mph (2,140 kph), more than twice that of conventional aircraft. Only 20 were ever built and the last Concorde flight was in 2003.

● Other industrial wonders

FOR THE CHANNEL TUNNEL IN 1802, 192 YEARS BEFORE IT OPENED.

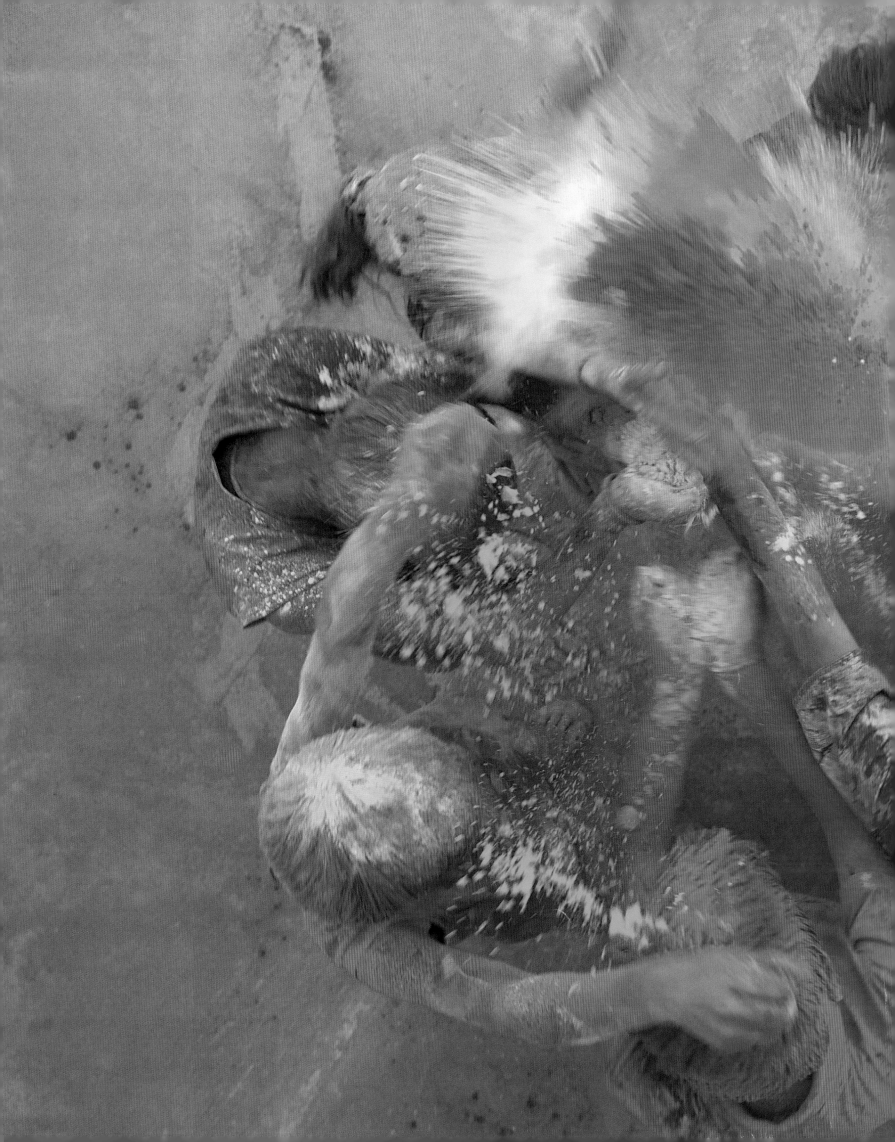

Culture

Holi Festival, Jodhpur, India
During the Hindu spring
festival of Holi—known as
the Festival of Colors—people
throw pigments and colored
water over each other.

Introduction

The word "culture" is a broad idea, and includes the values, beliefs, and behavior of a society, or group of people. Culture includes many things, including customs, language, religion, music, art, food, and clothing. Some points of culture are traditional, having survived virtually unchanged for centuries. Others are short-lived, such as fashion styles and trends in pop music.

Modern culture

Today's culture is fast-moving and ever-changing, thanks in part to the instant communication offered by the Internet. But long before the Internet, the migration of people around the world began introducing people to cultures different from their own. Global broadcasting then accelerated this effect in the 20th century. The cultural contact often creates a fusion (uniting) of different cultural styles, especially in the fields of music, fashion, and cooking.

Live performances
Huge crowds watch singers, such as Beyoncé (right), perform live, just as they have always done. But today the "live" audience can number many millions, with most following remotely via Internet-based platforms like YouTube or Spotify.

Stadium spectators
For many sports fans, being part of a passionate, noisy, banner-waving stadium crowd makes them feel an important part of the event.

Headdress, called a *kiritam*, varies in size and design, according to the character being portrayed.

Hand gestures (known as *mudra*) are the dancer's main way of telling the story.

Noble-hearted characters always have green faces; dark red signifies a treacherous nature.

Kathakali dancer
Indian kathakali dancers enact stories from two epic poems, the *Ramayana* and the *Mahabarata*. Dealing with the constant struggle between good and evil, dances end with the destruction of a demon.

Heroes always wear red jackets

Dancer's skirt is made up of many layers of white cotton.

DRAGON DANCES ARE TRADITIONAL PERFORMANCES IN CHINESE

Traditional culture

Older people can pass culture on to the next generation, enabling a society's traditions to be preserved for many years. The *Ramayana*, a Hindu poem written in the 5th or 4th century BCE, tells the story of Rama and Sita, and their battle against the demon-king Ravana. Over many generations, the *Ramayana* and its values have been kept alive in India and southern Asia through writing, story telling, painting, sculpture, festivals, music, and dance.

Literature
The *Ramayana* was originally written in Sanskrit, the language of Hinduism and ancient Indian literary texts.

Sculpture
The great warrior Rama, holding his bow, stands next to his wife Sita. Both hold up their right hands in blessing.

Festival
At the Hindu festival of Diwali, people light lamps to commemorate Rama's return from exile and his victory over Ravana.

Painting
In this scene from the *Ramakien*, a Thai version of the *Ramayana*, the monkey god Hanuman uses his body as a bridge for Rama to cross.

Music
Musicians in Bali, Indonesia, provide accompaniment to kecak dancers, who perform parts of the *Ramayana*.

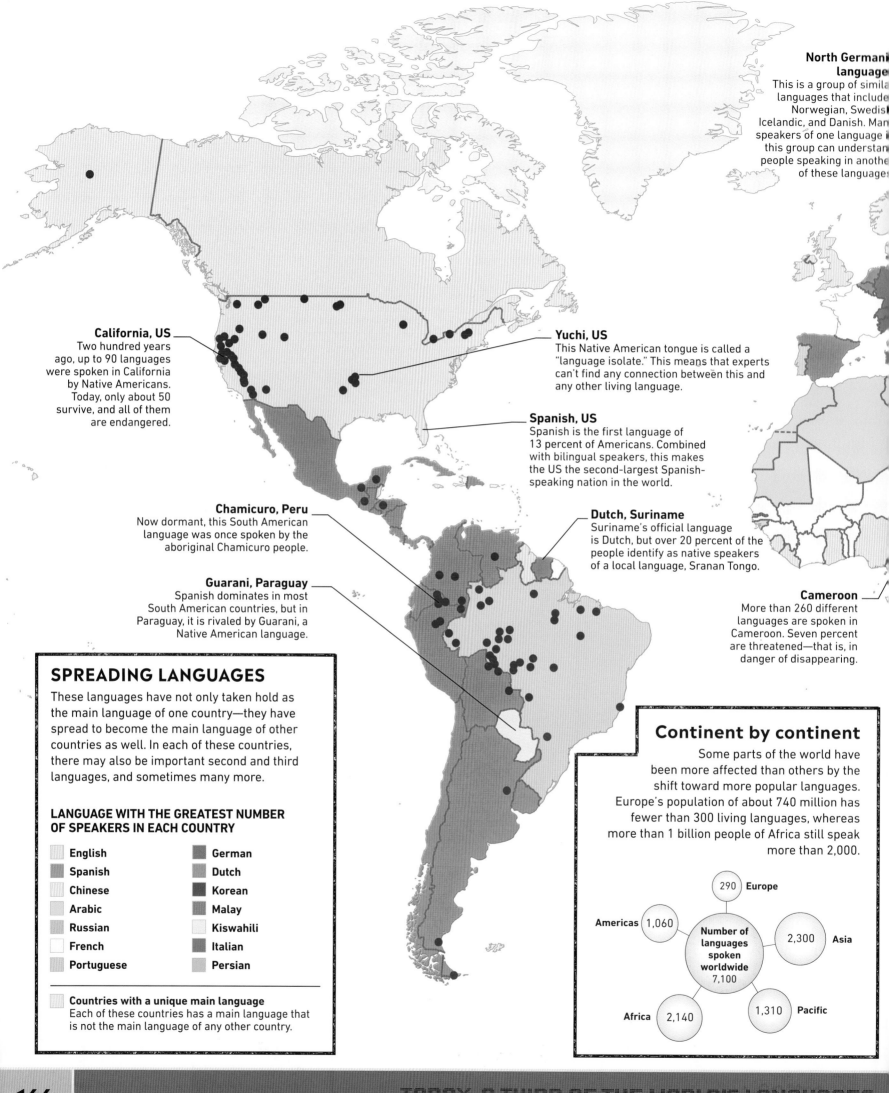

**North German
language**
This is a group of simila
languages that include
Norwegian, Swedis
Icelandic, and Danish. Man
speakers of one language i
this group can understan
people speaking in anothe
of these language

California, US
Two hundred years
ago, up to 90 languages
were spoken in California
by Native Americans.
Today, only about 50
survive, and all of them
are endangered.

Yuchi, US
This Native American tongue is called a
"language isolate." This means that experts
can't find any connection between this and
any other living language.

Spanish, US
Spanish is the first language of
13 percent of Americans. Combined
with bilingual speakers, this makes
the US the second-largest Spanish-
speaking nation in the world.

Chamicuro, Peru
Now dormant, this South American
language was once spoken by the
aboriginal Chamicuro people.

Dutch, Suriname
Suriname's official language
is Dutch, but over 20 percent of the
people identify as native speakers
of a local language, Sranan Tongo.

Cameroon
More than 260 different
languages are spoken in
Cameroon. Seven percent
are threatened—that is, in
danger of disappearing.

Guarani, Paraguay
Spanish dominates in most
South American countries, but in
Paraguay, it is rivaled by Guarani, a
Native American language.

SPREADING LANGUAGES

These languages have not only taken hold as
the main language of one country—they have
spread to become the main language of other
countries as well. In each of these countries,
there may also be important second and third
languages, and sometimes many more.

**LANGUAGE WITH THE GREATEST NUMBER
OF SPEAKERS IN EACH COUNTRY**

English		German
Spanish		Dutch
Chinese		Korean
Arabic		Malay
Russian		Kiswahili
French		Italian
Portuguese		Persian

Countries with a unique main language
Each of these countries has a main language that
is not the main language of any other country.

Continent by continent

Some parts of the world have
been more affected than others by the
shift toward more popular languages.
Europe's population of about 740 million has
fewer than 300 living languages, whereas
more than 1 billion people of Africa still speak
more than 2,000.

290 Europe

Americas 1,060

**Number of
languages
spoken
worldwide
7,100**

2,300 Asia

Africa 2,140

1,310 Pacific

Ter Saami, Russia
This is one of the languages spoken by the Sami, the indigenous people of the Arctic. Research in 2010 found that there are only two speakers left.

Russia
Apart from the main language—Russian—there are eight languages in Russia each with 1 million or more speakers, including Tatar, Ukrainian, Bashkir, Chechen, and Chuvash. At least 10 of Russia's minority languages are nearly extinct.

ON AVERAGE, **TWO** **LANGUAGES** DIE OUT **EVERY** **MONTH**

Chinese
About 1.3 billion people speak Mandarin Chinese. It has more native speakers than any other language.

India's languages
There are 22 official languages in India, although the national government uses only Hindi and English.

Papua New Guinea
About 800 languages are spoken here, making it the most varied place on Earth for languages.

ENDANGERED LANGUAGES
A language disappears when its speakers no longer exist. Factors such as war and urban growth can cause old languages to die out. Languages are quickly going extinct—more than 200 languages are spoken by fewer than 10 people each.

● **Languages with 10 or fewer native speakers**

Vanuatu
A total of 18 languages in Vanuatu now have fewer than 10 fluent speakers.

Languages

Languages were developed by humans so that they could communicate with each other within their groups. As communities began to interact more, some languages spread and became more widely spoken, whereas others were used less or even died out.

Australia's languages
There were about 250 indigenous Australian languages. Of the 120 that remain, well over half are endangered.

A place that religious followers think of as "holy" may be the spiritual center of the religion. It could be the place where it all began, a site of pilgrimage, or the religion's official headquarters.

Religious followers

Most of the world's people identify with a religion, whether or not they take part in religious services. Their beliefs, customs, or ancestors link them to their religious community.

Judaism	Sikhism	Buddhism	Hinduism	Islam	Christianity
14 million	30 million	535 million	1.2 billion	1.8 billion	2.5 billion

JERUSALEM

Contains sites holy to three major world religions.

✡ **Western Wall (Judaism)**
(1) Remains of the Temple in Jerusalem, and sacred site of prayer for Jews.

✝ **Church of the Holy Sepulchre**
(1) Said by Christians to contain the burial site of Jesus Christ.

☪ **Al-Aqsa Mosque (Islam)**
(1) The third-holiest place in Islam, where the Prophet Muhammad is said to have risen to heaven.

NEW FAITHS

New religions have emerged in the last 200 years.

✿ **BAHA'I, 1866**
(30) **Shrine of the Bab, Haifa**
Resting place of the Bab, revered by the Baha'i faith as a Messenger.

RASTAFARI, 1930
(31) **Jamaica**
Home of Rastafari, whose followers worship Haile Selassie I of Ethiopia as God in human form.

ISKCON, 1966
(32) **New York City**
The International Society for Krishna Consciousness, known as Hare Krishna, began here.

CHRISTIANITY

Followers worship Jesus Christ as the son of God. Christianity is split into these major branches: Orthodox, Catholic, and Protestant.

(2) **St. Mary of Zion Church**
Heart of the Ethiopian Orthodox Church, said to hold God's 10 Commandments in the Ark of the Covenant.

(3) **The Hand of God, Nigeria**
This megachurch can seat up to 120,000 people within its handlike layout.

(4) **Vatican City**
Headquarters of the Roman Catholic Church.

(5) **Our Lady of Guadelupe**
Mexico City's famous image of the Virgin Mary and site of a Roman Catholic pilgrimage.

ALMOST 220 MILLION HINDUS TRAVELED TO PRAYAGRAJ, INDIA, IN

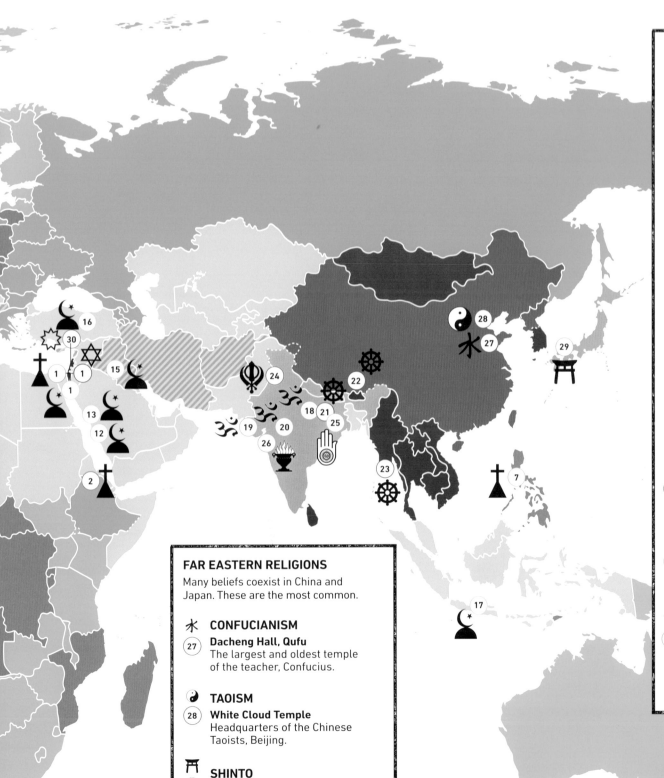

INDIAN RELIGIONS
Many world religions began in India, or, like Zoroastrianism, have taken up home there.

⚛ HINDUISM

18 **Varanasi**
Holiest Hindu city. Steps lead bathers down to the sacred River Ganges.

19 **Dwarka**
Pilgrimage site, holy city, and one of the Char Dam ("four seats") of Hinduism.

20 **Ujjain**
One of seven places (also including Dwarka and Varanasi) of "sacred ground."

☸ BUDDHISM

21 **Bodh Gaya**
Place where the Buddha, the founder of Buddhism, found enlightenment.

22 **Jokhang, Lhasa, Tibet**
The most important and sacred temple in Tibetan Buddhism.

23 **Shwedagon Pagoda**
In Yangon, Myanmar, this huge gold-plated building houses relics of the Buddha.

☬ SIKHISM

24 **Harmandir Sahib**
Known in English as the Golden Temple of Amritsar and sacred to Sikhs.

✋ JAINISM

25 **Pawapuri**
Sacred to the Jain faith, the site where a key teacher achieved enlightenment.

⚱ ZOROASTRIANISM

26 **Iranshah Atash Behram, Udvada, India**
An important fire temple of the Zoroastrian faith, which began in Persia (Iran).

FAR EASTERN RELIGIONS
Many beliefs coexist in China and Japan. These are the most common.

水 CONFUCIANISM
27 **Dacheng Hall, Qufu**
The largest and oldest temple of the teacher, Confucius.

☯ TAOISM
28 **White Cloud Temple**
Headquarters of the Chinese Taoists, Beijing.

⛩ SHINTO
29 **Izumo Taisha, Japan**
The Japanese emperor's family shrine.

---

6 **Our Lady of Aparecida, São Paulo, Brazil**
Eight million Catholic pilgrims a year visit this celebrated statue of the Virgin Mary.

7 **San Agustin Church, Manila**
The Philippines' oldest church, dating from 1607.

8 **All Saint's Church, Germany**
In Wittenberg, Martin Luther began Protestantism by nailing his ideas on the church door.

9 **Canterbury Cathedral**
Place of pilgrimage and world center of the Anglican Protestant Church.

10 **St. Peter's Church**
The oldest Anglican church outside Britain, in Bermuda.

11 **Salt Lake Temple**
Largest center of worship of the Church of Jesus Christ of Latter-day Saints, known as the Mormon Church.

☪ ISLAM
Muslims, followers of Islam, believe in one god and that Muhammad (570–632 CE) is His prophet. This religion split into Sunni and Shi'a faiths early on.

12 **Makkah**
Sacred to all Muslims as Muhammad's birthplace.

13 **Medinah**
The burial site of Islam's prophet, Muhammad.

14 **Kairouan, Tunisia**
Fourth city of Sunni Islam, and seat of Islamic learning.

15 **Najaf, Iraq**
Third city of Shi'a Muslims. Features the tomb of their first imam, Imam Ali.

16 **Konya, Turkey**
Home of Sufi mystic Rumi, whose followers perform the "Whirling Dervish" dance.

17 **Demak Great Mosque**
One of Indonesia's oldest mosques, built in the 15th century.

Raft the Salmon River, Idaho
Ride the rapids as you travel through spectacular canyons on the "River of No Return."

Surf at Mavericks, California
Only a select few are prepared to risk the big, wild waves at Mavericks, which can reach 50 ft (15 m).

Kauna'oa Bay, Hawaii
As well as swimming and sunbathing, you can snorkel to investigate local marine life. At night, you can even watch manta rays feeding in the bay.

Gran Cenote, Mexico
Divers can marvel at stalactites and stalagmites in this huge undersea cave formation.

Bora Bora, French Polynesia
Just 18 miles (29 km) long, this little island—the remnant of an extinct volcano—has beautiful white sandy beaches in a turquoise lagoon fringed by palm trees.

Galápagos Islands, Ecuador
These isolated islands boast many unique species, including giant tortoises, marine iguanas, and many different types of finches.

Palm Beach, Florida
Loved by millionaires, Palm Beach offers warm water, a fine climate, and bright city lights close at hand.

London Eye, UK
At its highest point, 443 ft (135 m) above the ground, the London Eye offers a panoramic view that stretches 25 miles (40 km) to the horizon.

Wiener Riesenrad, Vienna, Austria
Built in 1897, this 213-ft- (65-m-) tall structure was one of the first Ferris wheels ever made.

Bonaire, Caribbean
There are more than 80 superb dive sites around this small island, which is home to three species of sea turtle.

Hike the Inca Trail, Peru
Hike through mountains and jungles to the wonderfully preserved remains of the Inca city of Machu Picchu. There are strict visitor quotas, in an attempt to avoid damage to the 15th-century settlement.

Pantanal, Brazil/Bolivia/Paraguay, South America
The world's greatest concentration of jaguars—and much more besides. More than 1,000 bird species, including storks and macaws, and 300 types of mammals such as tapirs and anteaters.

FRANCE IS THE WORLD'S TOP TOURIST DESTINATION, WITH 89 MILLION VISITORS IN 2018

ABOUT 1.5 MILLION WILDEBEEST MIGRATE THROUGH KENYA'S MASAI

KEY

 Adventure destinations
These spots are for those who like their holidays thrill-packed, offering extreme activities such as white-water rafting, skydiving, surfing, and trekking in remote regions.

 World's top big wheels
Why not take a city break and ride one of the world's amazing observation wheels? Watch the world turn and take in the incredible views from the top.

 Best diving and snorkeling sites
Take the plunge and immerse yourself in the magical worlds of coral reefs and undersea caverns. Be careful not to touch the coral, though, as it's easily damaged.

Top 5 Beaches
Relax, stretch out, and catch some rays on a sandy shore somewhere. Can't decide where to go? No worries—we've done the hard work for you and picked the best of the bunch.

 Top 5 Safari sites
Get right up close to nature on a safari. See wild animals in their natural habitats, experience incredible animal migrations, and marvel at unique species.

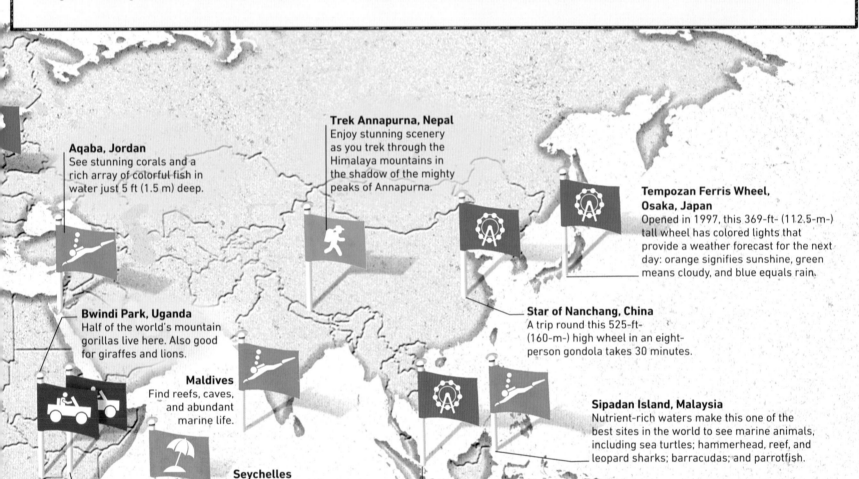

Aqaba, Jordan
See stunning corals and a rich array of colorful fish in water just 5 ft (1.5 m) deep.

Trek Annapurna, Nepal
Enjoy stunning scenery as you trek through the Himalaya mountains in the shadow of the mighty peaks of Annapurna.

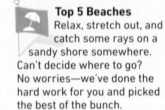 **Tempozan Ferris Wheel, Osaka, Japan**
Opened in 1997, this 369-ft- (112.5-m-) tall wheel has colored lights that provide a weather forecast for the next day: orange signifies sunshine, green means cloudy, and blue equals rain.

Bwindi Park, Uganda
Half of the world's mountain gorillas live here. Also good for giraffes and lions.

Star of Nanchang, China
A trip round this 525-ft- (160-m-) high wheel in an eight-person gondola takes 30 minutes.

Maldives
Find reefs, caves, and abundant marine life.

Sipadan Island, Malaysia
Nutrient-rich waters make this one of the best sites in the world to see marine animals, including sea turtles; hammerhead, reef, and leopard sharks; barracudas; and parrotfish.

Seychelles
Northeast of Madagascar, this beautiful archipelago is made up 155 islands.

Masai Mara, Kenya
See lions, leopards, and cheetahs, and the spectacular mass migration of zebras, gazelles, and wildebeest.

Singapore Flyer
One of the world's tallest observation wheels, at 541 ft (165 m), which gives views of 28 miles (45 km).

Okavango Delta, Botswana
Watch large roaming herds of buffaloes and elephants, and endangered animals such as African wild dogs.

 Fraser Island, Australia
This World Heritage Site has 640 sq miles (1,660 sq km) of unspoiled natural beauty.

Tourism

Traveling can offer adventure, fun, and an unforgettable glimpse of the world's natural wonders—but it's important to consider the environmental impact of tourism, too. In 2020, the industry was severely affected by the COVID-19 pandemic.

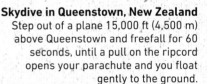 **Skydive in Queenstown, New Zealand**
Step out of a plane 15,000 ft (4,500 m) above Queenstown and freefall for 60 seconds, until a pull on the ripcord opens your parachute and you float gently to the ground.

Edward Hopper
1882–1967; US.
Hopper painted in the Realist style, which tries to show things as they are in real life. Hopper used simple colors and often painted solitary, lonely-looking people.

Andy Warhol
1928–1987; US.
Warhol pioneered Pop Art—the "pop" refers to popular culture. His art used familiar images of famous people and everyday items such as soup cans. Warhol took his inspiration from advertising, TV, and comic strips.

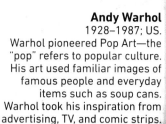

Edvard Munch
1863–1944; Norway.
Munch was an Expressionist artist. Expressionists tried to express feelings in their work, rather than portray people and objects accurately. Munch's most famous painting is *The Scream* (1893), which shows a person with an agonized expression.

Thomas Gainsborough
1727–88; England.
Founder of the 18th-century British Landscape school, Gainsborough also made portraits. *Mr. and Mrs. Andrews* (1750; right) is an early masterpiece.

Claude Monet
1840–1926; France.
Impressionists such as Monet painted their view of brief moments in time.

Frida Kahlo
1907–1954; Mexico.
Frida Kahlo began painting after she was badly injured in an accident. She is best known for her self-portraits. Her work used bold, bright colors and was influenced by Mexican folk art.

Pablo Picasso
1881–1973; Spain.
Among many other things, this famous artist was a founder of Cubism—a style that used shapes to depict people and objects, often showing them from multiple viewpoints at the same time.

Victor Meirelles
1832–1903; Brazil.
Meirelles' religious and military paintings and depictions of episodes from Brazilian history won him fame and praise in the 19th century. His painting *The First Mass in Brazil* (1860; right) still appears in primary-school history books in Brazil.

Eugène Delacroix
1798–1863; France.
Delacroix was one of the Romantics, who stressed imagination and emotion. *Liberty Leading the People* (1830; above) marks the overthrow of Charles X of France in 1830.

Sculpture
13th century–present; Nigeria.
The people of the Kingdom of Benin, in what is now Nigeria, sculptured bronze heads and figures. They also made masks out of wood, bronze, and ivory. The tradition continues: on the right is a wooden mask of the late 20th century.

Art

People the world over value art because it allows them to express their emotions and their culture, record history and everyday life, and explore what it means to be human. The works of the world's great artists often sell for huge sums of money.

Marc Chagall
1887–1985; Russia.
Chagall produced Expressionist and Cubist paintings, and also stained-glass windows. He is known for his paintings of village scenes and of lovers floating in the air.

Yue Minjun
Born 1962; China.
Based in Beijing, Yue Minjun is best known for his oil paintings, which show him frozen with laughter in various poses and in different settings. He has also represented himself in sculptures, watercolor paintings, and prints. He first exhibited his work in 1987; by 2007, he had sold 13 paintings for more than $1 million each.

Tamara de Lempicka
1898–1980; Poland.
In the 1920s and 1930s, de Lempicka was the most famous painter in the Art Deco style, which featured geometric shapes and intense, bright colors. She lived a flamboyant life and associated with the rich and famous.

Caravaggio
1571–1610; Italy.
Caravaggio was one of the Baroque artists, who revolutionized art by painting realistic rather than idealized people and scenes. He is one of the most influential painters in art history.

Katsushika Hokusai
1760–1849; Japan.
Hokusai is perhaps the most famous Japanese printmaker. His wood-block prints included seascapes, such as *The Great Wave off Kanagawa* (1831; above), and scenes from everyday life.

Basawan
c.1580–1600; India.
A painter of miniature scenes, Basawan illustrated the *Akbarnama* (right)—the official chronicle of Akbar, the third Mughal Emperor.

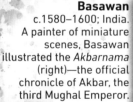

Willie Bester
Born 1956; South Africa.
Bester's collages and sculptures use recycled material and objects found in scrapyards and flea markets. His 1992 *Tribute to Biko* (above) commemorates Stephen Biko, who campaigned for racial equality in South Africa.

Yannima Tommy Watson
1935–2017; Australia.
Despite starting painting only in 2001, when he was in his mid-60s, Tommy Watson rapidly became one of Australia's foremost Aboriginal artists. His paintings relate to the stories of the Dreamtime—the creation period in Aboriginal mythology.

IT IS ESTIMATED THAT **PICASSO** PRODUCED ABOUT **148,000** **WORKS OF ART** DURING HIS **LIFETIME**

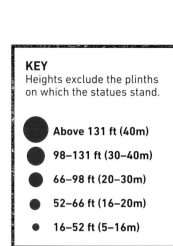

Christ the King
120 ft (36 m)
Swiebodzin, Poland
2010

4. Statue of Liberty
Liberty was a gift from the people of France to the US.

Angel of the North
66 ft (20 m) high,
175 ft (54 m) wingspan
Gateshead, UK
1998

THE **ANGEL OF THE NORTH** HAS A BIGGER WINGSPAN THAN A **BOEING 767** JET

Moai (statues)
Up to 33 ft (10 m)
Easter Island
1100 CE–1650 CE

Christ the Redeemer
98 ft (30 m)
Rio de Janeiro,
Brazil, 1931

Great Sphinx
66 ft (20 m)
Giza, Egypt
2500 BCE

Statues

Since ancient times, humans have built grand statues of great rulers, heroic figures, and gods and goddesses. We are still doing it, and statues today are getting bigger and bigger.

Political statues

Some statues are built to remind people of their freedoms, promote a sense of unity, or reinforce political ideas.

1. The Motherland Calls
279 ft (85 m); Volgograd, Russia; 1967
Marks the Soviet Union's victory over German forces in the Battle of Stalingrad (1942–43).

2. Mother of the Fatherland
203 ft (62 m); Kiev, Ukraine; 1981
The female statue represents the strength and victory of the Soviet Union in World War II.

3. African Renaissance Monument
161 ft (49 m); Dakar, Senegal; 2010
Africa's tallest statue shows a man gazing out to sea as he holds a woman and child.

4. Statue of Liberty
151 ft (46 m); New York, US; 1886
"Lady Liberty" stands with a torch in one hand and a stone tablet in the other.

5. Juche Tower statues
98 ft (30 m); Pyongyang, North Korea; 1982
Three figures represent a peasant, an industrial worker, and an intellectual.

African Renaissance Monument

THE STATUE OF LIBERTY IN NEW YORK CONTAINS

1. The Motherland Calls
The statue beckons fighters to come to the defense of their nation.

The Statue of Unity
597 ft (182 m)
Gujarat, India; 2018
The world's tallest statue, depicting India's first Deputy Prime Minister.

Spring Temple Buddha
420 ft (128 m)
Lushan, China; 2002
Named after the nearby Tianrui hot spring.

Religious statues

Many religious movements use statues to inspire belief and to aid worship.

Guanyin, Hainan, China

11. Buddha
381 ft (116 m); Monywa, Myanmar; 2008
Depicts the Buddha standing. World's third-tallest statue.

12. Guanyin
354 ft (108 m); Sanya, Hainan, China; 2005
Represents the goddess Guanyin blessing the world.

13. Virgin of Peace
154 ft (47 m) Trujillo, Venezuela; 1983
The Virgin Mary, mother of Jesus, is shown holding a dove of peace in her hand.

14. Shiva
143 ft (44 m); Chitapol, Kathmandu, Nepal; 2012
Hindu god Shiva stands with a trident in his left hand. His right hand offers a blessing.

15. Murugan
141 ft (43 m); Batu Caves, Gombak, Malaysia; 2006
Statue stands by a cave shrine to the Hindu god Murugan.

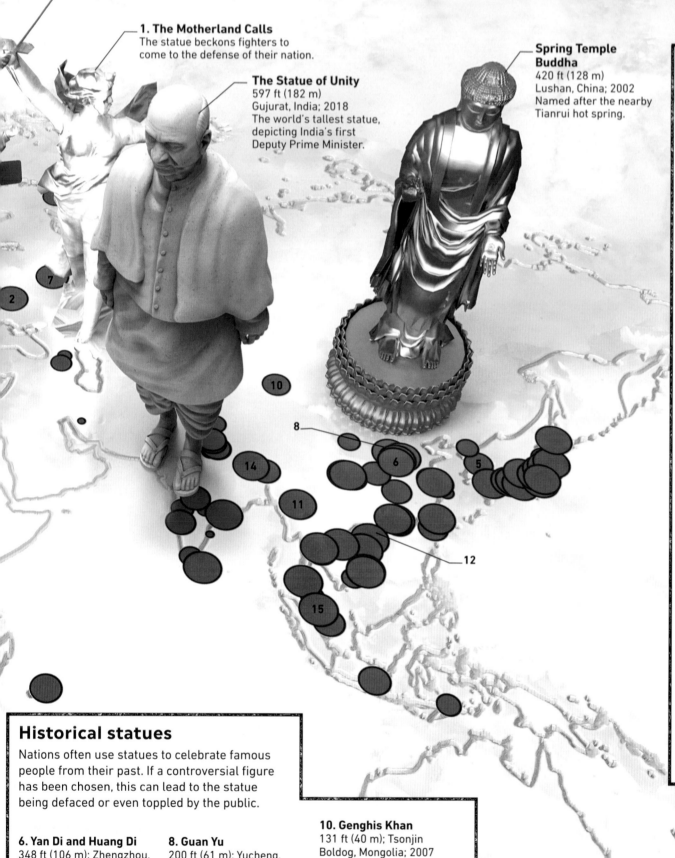

Historical statues

Nations often use statues to celebrate famous people from their past. If a controversial figure has been chosen, this can lead to the statue being defaced or even toppled by the public.

6. Yan Di and Huang Di
348 ft (106 m); Zhengzhou, China; 2007
Shows the heads of two legendary kings regarded as the early founders of the Chinese nation.

7. Peter the Great
315 ft (96 m); Moscow, Russia; 1997
Erected to celebrate 300 years of the Russian Navy, which Tsar Peter I founded.

8. Guan Yu
200 ft (61 m); Yucheng, Shanxi, China; 2010
Statue of the general Guan Yu (160–219), later deified as Chinese god of war, at his birthplace.

9. José Maria Morelos
131 ft (40 m); Janitzio, Michoacán, Mexico; 1934
Mexico's rebel leader in the War of Independence (1810–21), fist clenched.

10. Genghis Khan
131 ft (40 m); Tsonjin Boldog, Mongolia; 2007
This statue depicts the famous Mongol leader (ruled 1206–1227) mounted on a horse.

Peter the Great

Tulip Time Festival
Michigan
This festival is held in cities that were founded by the Dutch or had large numbers of Dutch settlers. Tulips line the streets and special tulip gardens are created for the event.

Vancouver, Canada

San Francisco, California

Thanksgiving
US and Canada
This harvest celebration in November (October in Canada) usually involves a turkey dinner. It was first held to give thanks for the harvest of 1621.

Toronto, Canada

New York, New York

Mardi Gras, New Orleans, Louisiana

Noche de Rábanos
Oaxaca, Mexico
When radishes first were brought to the Americas in the 16th century, market traders made radish sculptures to advertise the new vegetables. The "Night of the Radishes" has celebrated that custom since 1897.

Cuba

Haiti

Jamaica

Barranquilla, Colombia

Mazatenango, Guatemala

Panama

Trinidad

Tapati Festival
Easter Island
Tapati includes dancing, ritual chants, art exhibits, carving competitions, horse and boat races, body-painting, a string figure (*kai-kai*) contest, the selection of a queen, a parade, and *haka pei*—sliding down a steep hillside on banana-tree trunks at high speed.

Ambato, Ecuador

French Guiana

Cajamarca, Peru

Inti Raymi Day
Cuzco, Peru
The Festival of the Sun dates back to the Incas. People celebrate the winter solstice and the start of the new year.

Oruro, Bolivia

Coastal cities, Brazil

Montevideo, Uruguay

Cologne, Germany

Binche, Belgium

Notting Hill London, UK

Cheese Rolling Festival
Gloucestershire, UK
Contestants chase a wheel of cheese down a steep, muddy hill.

Venice, Ita

Tomatina
Buñol, Valencia
Since 1944, tomato fights have been held on the last Wednesday of August. More than 110 tons of tomatoes are hurled each year!

Ovar, Portugal

Malt

Festival of the Sahara
Tunisia
A festival celebrating nomadic life and traditions. Events include camel marathons and performances of Bedouin song, dance, and poetry.

Madeira

Santa Cruz, Tenerife

Cape Verde Islands

Festival-au-Desert
Mali
Three days of traditional Tuareg art, music, and dance. Everyone camps in the desert, with their camels close by.

Festivals

Festivals give people a chance to celebrate their religious and cultural traditions. Above all, they are a great opportunity to throw a party!

Carnival

Carnival is marked by parades, such as in Rio de Janeiro, Brazil (left). It comes just before Lent—a time of fasting and avoiding rich foods that leads up to the Christian festival of Easter.

Major Carnival locations

Wife-Carrying World Championships
Sonkajärvi, Finland
Male entrants carry their wives over an obstacle course. The winner receives his wife's weight in beer.

Baltai
Tatarstan, Russia
Baltai means "feast of honey." The festival marks the start of the mowing season and is celebrated by decorating a bear with birch leaves.

Chinese New Year

Called the Spring Festival in China, since it marks the end of winter, this festival typically involves street processions with lanterns and Chinese dragons. Families clean their houses to sweep away bad fortune and welcome in the New Year. The festival is celebrated in all countries with significant populations of Chinese people.

 Locations with important Chinese New Year celebrations

Rijeka, Croatia

Patras, Greece

Limassol, Cyprus

Beijing

Ghost Festival, *China*
Part of "Ghost Month," when the ghosts and spirits of dead ancestors are said to emerge from the underworld.

Kolkata, India

Janmashtami
Mumbai, India
Marks the birthday of the Hindu god Krishna. Boys and men clamber to the top of a pole, trying to smash a clay pot full of curd and spill its contents. Krishna is said to have stolen curd from pots as a boy.

Goa, India

Boryeong Mud Festival
Boryeong, South Korea
At this mucky festival, which dates from 1998, people cover each other in mud. The mud is said to contain minerals that are good for the skin.

Asakusa district, Tokyo

Awa Odori
Tokushima, Japan
Awa Odori began in 1586, when Tokushima's residents decided to celebrate their town's new castle. Today, more than 1 million tourists visit to watch performers in traditional dress dance in the streets.

Philippines

Singapore

Indonesia

Esala Maha Perahera
Kandy, Sri Lanka
The 10-day "Festival of the Tooth" celebrates the Tooth Relic of the Lord Buddha. Dancers, acrobats, and fire performers gather in Kandy. On the last night, an elegantly dressed elephant carries the tooth.

Mauritius

Bendigo Easter Festival
Bendigo, Australia
Dating from 1871, this is Australia's longest continuously running festival. During the festival's Easter procession, the *Sun Loong*, the longest imperial dragon in the world, dances through the streets of Bendigo.

Sydney, Australia

Incwala
Eswatini
At the "Festival of the first fruits," the king eats pumpkins and other fruits. People dance and sing in his honor and to bring blessings on the harvest.

World parties

Some festivals draw people from far and wide. They may be messy, such as Tomatina (left), or involve unusual competitions, such as wife-carrying.

 Key world party sites

Prickly Pear Festival
Mandela Bay, South Africa
This is a day for celebrating (and eating!) traditional foods such as ginger beer, pancakes, potjiekos, bunnychow, and fish braai.

Te Matatini
New Zealand
A Māori dance festival in which performers come together from all over New Zealand to compete in the national finals. *Te Matatini* means "many faces."

Television

Televisions provide us with entertainment and news 24 hours a day. People can also watch content on mobile devices such as laptops, smartphones, and tablets.

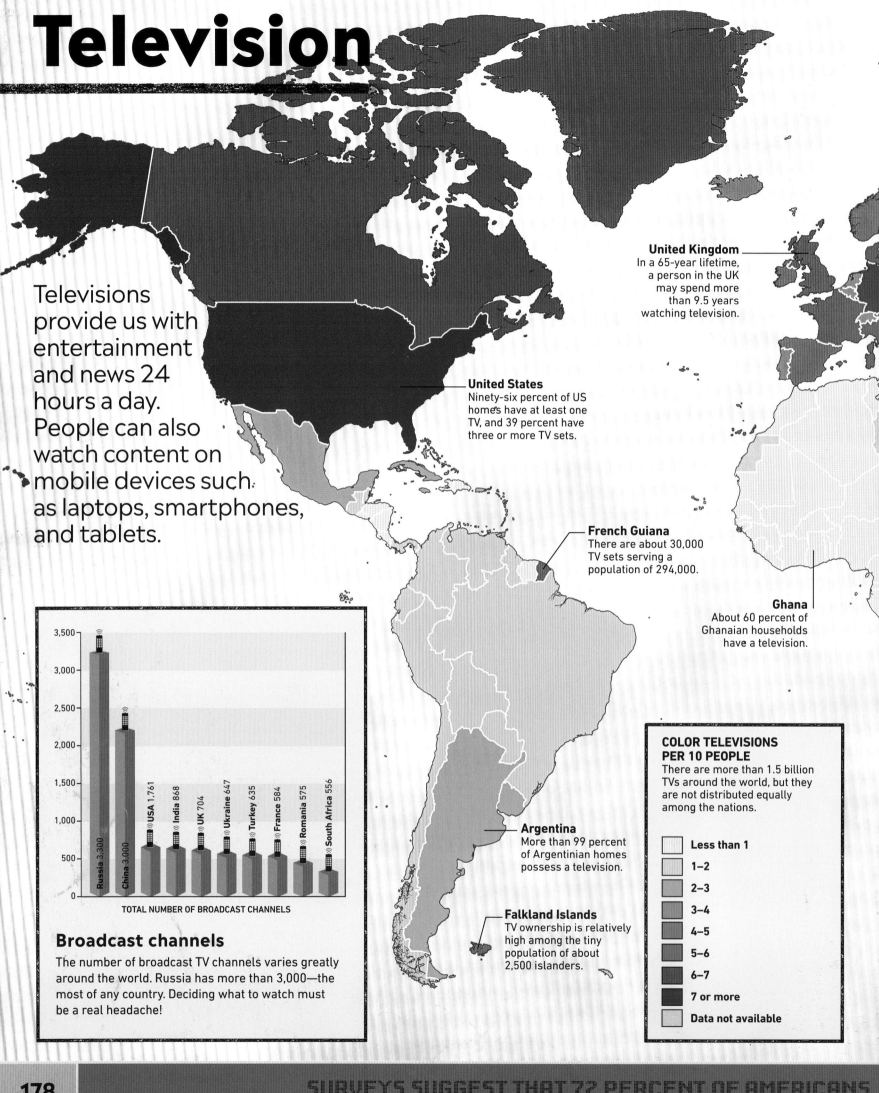

United Kingdom
In a 65-year lifetime, a person in the UK may spend more than 9.5 years watching television.

United States
Ninety-six percent of US homes have at least one TV, and 39 percent have three or more TV sets.

French Guiana
There are about 30,000 TV sets serving a population of 294,000.

Ghana
About 60 percent of Ghanaian households have a television.

Argentina
More than 99 percent of Argentinian homes possess a television.

Falkland Islands
TV ownership is relatively high among the tiny population of about 2,500 islanders.

Broadcast channels

The number of broadcast TV channels varies greatly around the world. Russia has more than 3,000—the most of any country. Deciding what to watch must be a real headache!

TOTAL NUMBER OF BROADCAST CHANNELS

- Russia 3,300
- China 3,000
- USA 1,761
- India 868
- UK 704
- Ukraine 647
- Turkey 635
- France 584
- Romania 575
- South Africa 556

COLOR TELEVISIONS PER 10 PEOPLE
There are more than 1.5 billion TVs around the world, but they are not distributed equally among the nations.

- Less than 1
- 1–2
- 2–3
- 3–4
- 4–5
- 5–6
- 6–7
- 7 or more
- Data not available

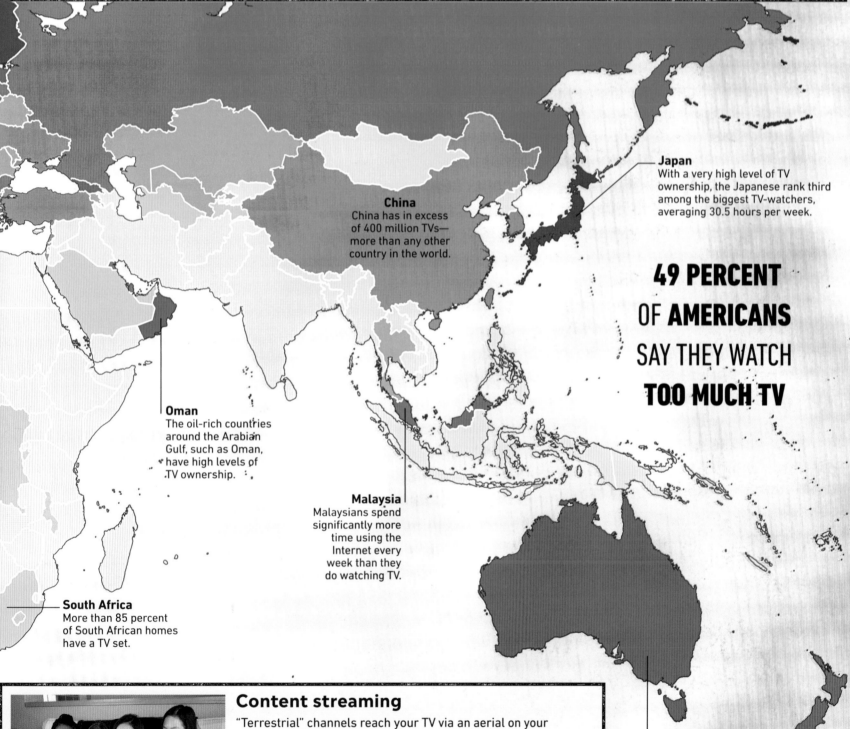

Hours per week

Experts say that watching more than 2 hours of TV per day (14 hours per week) can be bad for your health, yet in many countries, people watch twice that.

United States	Poland	Japan	Brazil	Russia	Italy	Spain	France	Germany	United Kingdom
31.5	30.8	30.5	29.6	28.9	28.9	27.2	26	26	24.7

HOURS PER PERSON PER WEEK

Japan
With a very high level of TV ownership, the Japanese rank third among the biggest TV-watchers, averaging 30.5 hours per week.

China
China has in excess of 400 million TVs—more than any other country in the world.

49 PERCENT OF **AMERICANS** SAY THEY WATCH **TOO MUCH TV**

Oman
The oil-rich countries around the Arabian Gulf, such as Oman, have high levels of TV ownership.

Malaysia
Malaysians spend significantly more time using the Internet every week than they do watching TV.

South Africa
More than 85 percent of South African homes have a TV set.

Australia
In 2017, Australian homes had an average of 6.4 screens per household.

Content streaming

"Terrestrial" channels reach your TV via an aerial on your home, while extra channels can be broadcast by satellite or sent through cables. Paying for cable TV has become steadily less popular with the rise of television streaming services such as Netflix, however, which involve playing video content over an Internet connection. Since the content isn't live, viewers can choose exactly what they want to watch, and when. In 2020, the streaming subscription market grew by a massive 37 percent.

Americas

1 Los Angeles Memorial Coliseum
California, US. Capacity 93,607; opened 1921

2 Rose Bowl
Pasadena, California, US. Capacity 92,542; opened 1922

3 Dodgers Stadium
California, US. Capacity 56,000; opened 1962

4 Estadio Monumental "U"
Lima, Peru. Capacity 80,093; opened 2000

5 Bell Center
Montreal, Canada. Capacity 21,273; opened 1996

6 Beaver Stadium
Pennsylvania, US. Capacity 106,572; opened 1960

7 Madison Square Garden
New York, US. Capacity 22,292; opened 1968

8 Arthur Ashe Stadium
New York, US. Capacity 23,200; opened 1997

9 Ohio Stadium
Ohio, US. Capacity 102,329; opened 1922

10 Neyland Stadium
Tennessee, US. Capacity 102,455; opened 1921

11 Sanford Stadium
Georgia, US. Capacity 92,746; opened 1929

12 Bryant–Denny Stadium
Alabama, US. Capacity 101,821; opened 1929

13 Tiger Stadium
Louisiana, US. Capacity 92,542; opened 1924

14 Darrell K. Royal—Texas Memorial Stadium Texas, US. Capacity 100,119; opened 1924

Michigan Stadium
Ann Arbor, Michigan. Capacity 114,804; opened 1926. Nicknamed "The Big House," this is the largest stadium in the US. It is home to the Michigan Wolverines American football team.

Camp Nou
Barcelona, Spain. Capacity 99,354; opened 1957. The largest stadium in Europe and 12th largest in the world.

KEY
The colors show capacity (numbers of spectators).

- 110,000 and above
- 100,000–109,999
- 90,000–99,999
- 80,000–89,999
- Fewer than 80,000

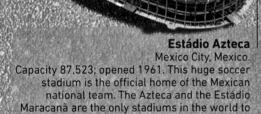

Estádio Azteca
Mexico City, Mexico. Capacity 87,523; opened 1961. This huge soccer stadium is the official home of the Mexican national team. The Azteca and the Estádio Maracanã are the only stadiums in the world to have hosted two FIFA World Cup soccer finals.

Estádio do Maracanã
Rio de Janeiro, Brazil. Capacity 82,238; opened 1950. Built for the 1950 football FIFA World Cup, the Maracanã was the world's largest stadium at the time, with room for nearly 200,000 people. Capacity was greatly reduced in the 1990s after part of the stadium collapsed. It served as the venue for the opening and closing ceremonies of the 2016 Summer Olympics and Paralympics.

Stadiums

Stadiums and arenas are among the largest and most impressive buildings on the planet. They not only enable us to experience the thrills and drama of competition between the best sports players, teams, and athletes, but also host pop concerts and other shows.

Europe

15 Millennium Stadium
Cardiff, UK. Capacity 74,500; opened 1999

16 Wembley Stadium
London, UK. Capacity 90,000; opened 2007

17 Allianz Arena
Munich, Germany. Capacity 69,901; opened 2005

18 Estádio Santiago Bernabéu
Madrid, Spain. Capacity 85,454; opened 1947

THE LARGEST EVER "MEXICAN WAVE" INVOLVED 157,574 PEOPLE A

THE RECORD FOR THE LOUDEST CROWD ROAR OF
142.2 DECIBELS WAS SET AT ARROWHEAD STADIUM,
KANSAS CITY, MISSOURI, DURING A FOOTBALL GAME IN 2014

Rungrado May Day Stadium
Pyongyang, North Korea.
Capacity 150,000; opened 1989.
Said to look like a magnolia
blossom, the stadium is used
for sports and military parades.

22 **23**

19

20

21

Record crowd sizes

Crowds were even larger before the
modern safety-conscious era, and
standing and overcrowding were
common. The largest-ever crowds
at sports events are below.

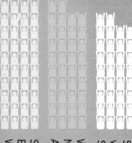

Soccer: 199,854. Maracanã Stadium, Brazil. Brazil vs Uruguay. World Cup Final. July 1950.

Wrestling: 190,000. May Day Stadium, North Korea. Pro-Wrestling event.

Soccer: 149,415 (plus 20,000 without tickets). Hampden Park, Scotland. Scotland vs England, 1937. April 1995.

Soccer: 135,000. Estádio da Luz, Portugal. Benfica vs Porto. January 1987.

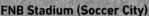

FNB Stadium (Soccer City)
Johannesburg, South Africa.
Capacity 94,736; opened 1989.
Nicknamed "The Calabash" because it looks
like the African pot of the same name, the FNB
is the largest stadium in Africa. The stadium
played host to the 2010 FIFA World Cup.

Melbourne Cricket Ground
Victoria, Australia.
Capacity 100,018; opened 1854.
This stadium holds the record
for the highest floodlight towers
of any sporting venue. It is
known to locals as "The G."

Asia

19 Azadi Stadium
Tehran, Iran. Capacity
100,000; opened 1971

20 Salt Lake Stadium
Kolkata, India. Capacity
120,000; built 1984

**21 Lumpinee
Boxing Stadium**
Bangkok, Thailand. Capacity
9,500; opened 1956

**22 Beijing National
Stadium ("Bird's Nest")**
China. Capacity: 80,000;
opened 2008

**23 Gwangmyeong
Velodrome**
South Korea.
Capacity 30,000;
opened 2006

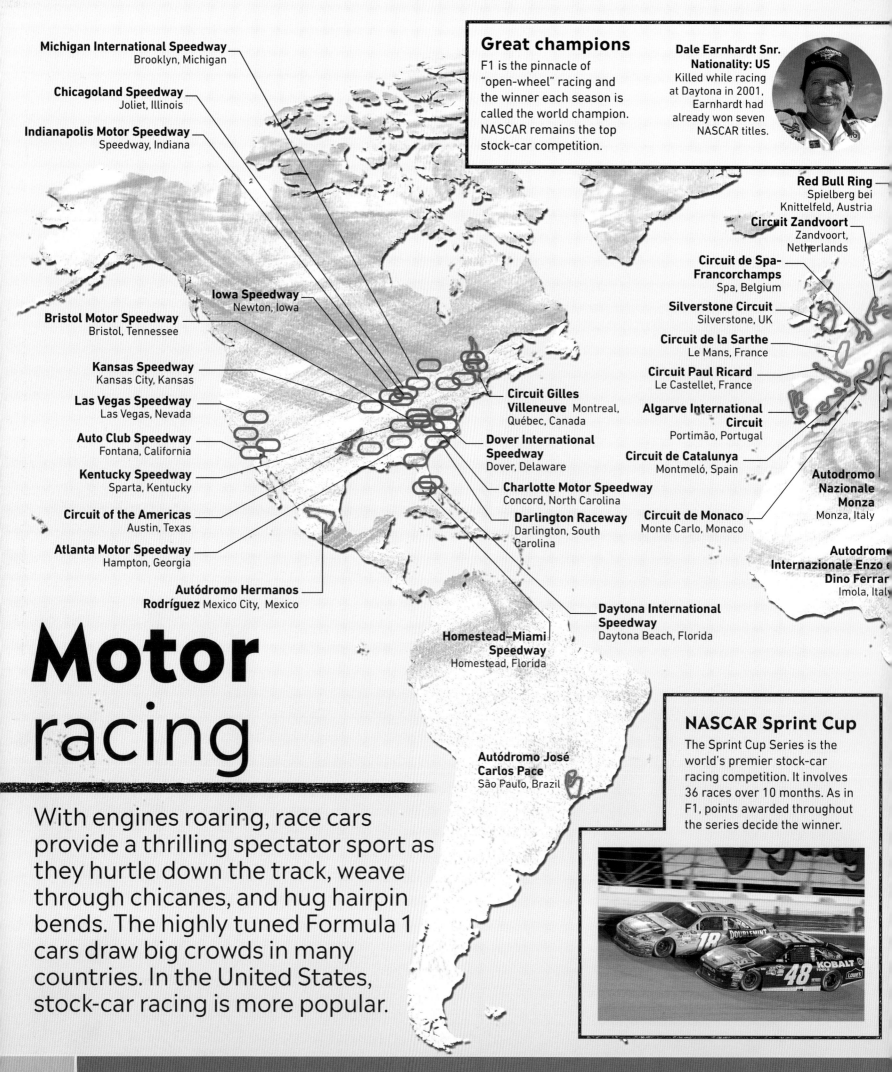

Michigan International Speedway
Brooklyn, Michigan

Chicagoland Speedway
Joliet, Illinois

Indianapolis Motor Speedway
Speedway, Indiana

Iowa Speedway
Newton, Iowa

Bristol Motor Speedway
Bristol, Tennessee

Kansas Speedway
Kansas City, Kansas

Las Vegas Speedway
Las Vegas, Nevada

Auto Club Speedway
Fontana, California

Kentucky Speedway
Sparta, Kentucky

Circuit of the Americas
Austin, Texas

Atlanta Motor Speedway
Hampton, Georgia

Autódromo Hermanos Rodríguez Mexico City, Mexico

Circuit Gilles Villeneuve Montreal, Québec, Canada

Dover International Speedway
Dover, Delaware

Charlotte Motor Speedway
Concord, North Carolina

Darlington Raceway
Darlington, South Carolina

Daytona International Speedway
Daytona Beach, Florida

Homestead–Miami Speedway
Homestead, Florida

Autódromo José Carlos Pace
São Paulo, Brazil

Red Bull Ring
Spielberg bei Knittelfeld, Austria

Circuit Zandvoort
Zandvoort, Netherlands

Circuit de Spa-Francorchamps
Spa, Belgium

Silverstone Circuit
Silverstone, UK

Circuit de la Sarthe
Le Mans, France

Circuit Paul Ricard
Le Castellet, France

Algarve International Circuit
Portimão, Portugal

Circuit de Catalunya
Montmeló, Spain

Circuit de Monaco
Monte Carlo, Monaco

Autodromo Nazionale Monza
Monza, Italy

Autodrome Internazionale Enzo e Dino Ferrar
Imola, Italy

Great champions

F1 is the pinnacle of "open-wheel" racing and the winner each season is called the world champion. NASCAR remains the top stock-car competition.

**Dale Earnhardt Snr.
Nationality: US**
Killed while racing at Daytona in 2001, Earnhardt had already won seven NASCAR titles.

Motor racing

With engines roaring, race cars provide a thrilling spectator sport as they hurtle down the track, weave through chicanes, and hug hairpin bends. The highly tuned Formula 1 cars draw big crowds in many countries. In the United States, stock-car racing is more popular.

NASCAR Sprint Cup

The Sprint Cup Series is the world's premier stock-car racing competition. It involves 36 races over 10 months. As in F1, points awarded throughout the series decide the winner.

ORDINARY CAR TYRES HAVE A LIFE OF 16,000 KM (10,000 MILES)

Michael Schumacher
Nationality: German
Seven-time F1 World Champion with 91 Grand Prix wins. He suffered a severe skiing accident in 2013 and has been receiving treatment ever since.

Ayrton Senna
Nationality: Brazilian
Three-time F1 World Champion. Fifth-most-successful driver of all time in terms of F1 race wins (41). Died in an accident at the 1994 San Marino Grand Prix.

Lewis Hamilton
Nationality: British
Jointly tied with Shumacher for the most World Championship titles, and holds the record outright for the most ever F1 wins.

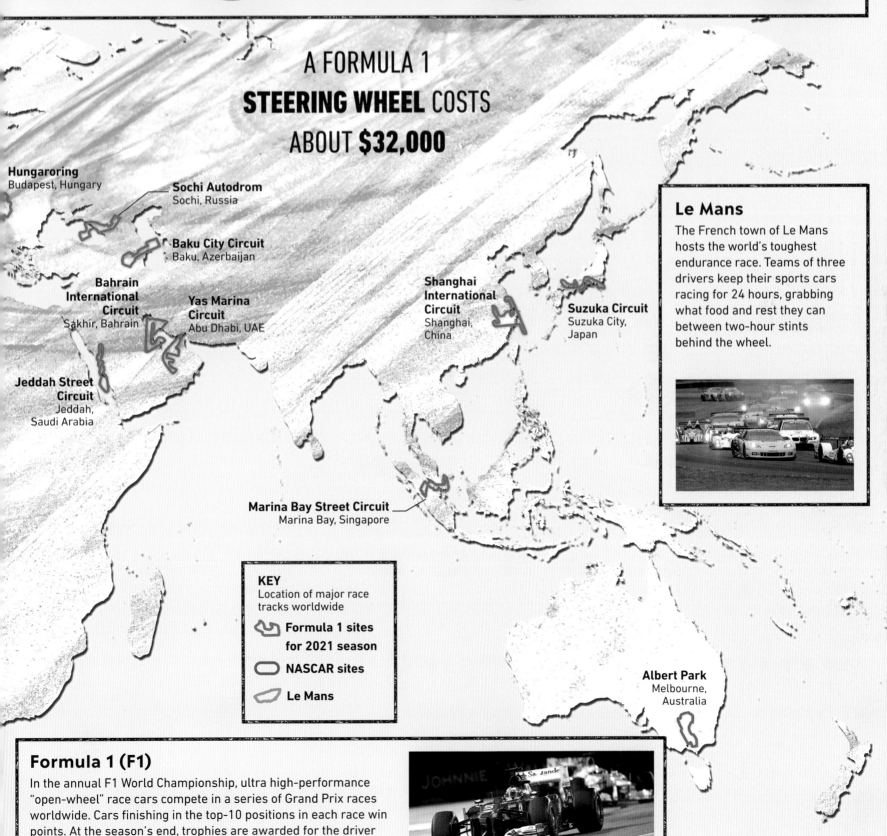

A FORMULA 1 STEERING WHEEL COSTS ABOUT $32,000

Hungaroring
Budapest, Hungary

Sochi Autodrom
Sochi, Russia

Baku City Circuit
Baku, Azerbaijan

Bahrain International Circuit
Sakhir, Bahrain

Yas Marina Circuit
Abu Dhabi, UAE

Shanghai International Circuit
Shanghai, China

Suzuka Circuit
Suzuka City, Japan

Jeddah Street Circuit
Jeddah, Saudi Arabia

Marina Bay Street Circuit
Marina Bay, Singapore

Albert Park
Melbourne, Australia

Le Mans

The French town of Le Mans hosts the world's toughest endurance race. Teams of three drivers keep their sports cars racing for 24 hours, grabbing what food and rest they can between two-hour stints behind the wheel.

KEY
Location of major race tracks worldwide

- Formula 1 sites for 2021 season
- NASCAR sites
- Le Mans

Formula 1 (F1)

In the annual F1 World Championship, ultra high-performance "open-wheel" race cars compete in a series of Grand Prix races worldwide. Cars finishing in the top-10 positions in each race win points. At the season's end, trophies are awarded for the driver and manufacturer with the most points.

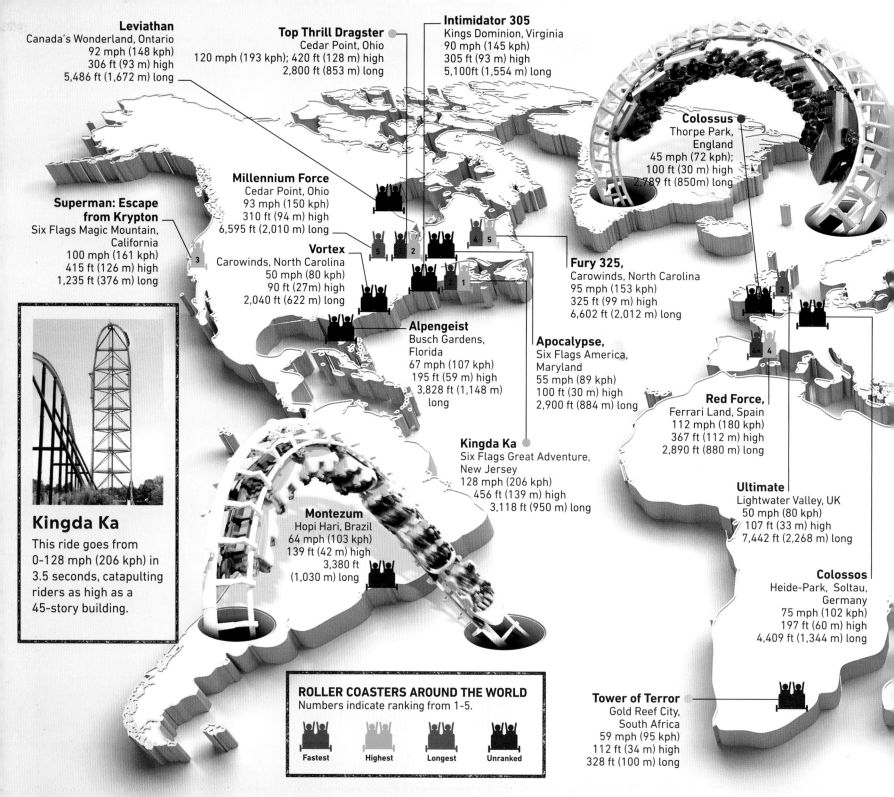

Leviathan
Canada's Wonderland, Ontario
92 mph (148 kph)
306 ft (93 m) high
5,486 ft (1,672 m) long

Top Thrill Dragster
Cedar Point, Ohio
120 mph (193 kph); 420 ft (128 m) high
2,800 ft (853 m) long

Intimidator 305
Kings Dominion, Virginia
90 mph (145 kph)
305 ft (93 m) high
5,100ft (1,554 m) long

Colossus
Thorpe Park,
England
45 mph (72 kph);
100 ft (30 m) high
2,789 ft (850m) long

Superman: Escape from Krypton
Six Flags Magic Mountain,
California
100 mph (161 kph)
415 ft (126 m) high
1,235 ft (376 m) long

Millennium Force
Cedar Point, Ohio
93 mph (150 kph)
310 ft (94 m) high
6,595 ft (2,010 m) long

Vortex
Carowinds, North Carolina
50 mph (80 kph)
90 ft (27m) high
2,040 ft (622 m) long

Fury 325,
Carowinds, North Carolina
95 mph (153 kph)
325 ft (99 m) high
6,602 ft (2,012 m) long

Alpengeist
Busch Gardens,
Florida
67 mph (107 kph)
195 ft (59 m) high
3,828 ft (1,148 m)
long

Apocalypse,
Six Flags America,
Maryland
55 mph (89 kph)
100 ft (30 m) high
2,900 ft (884 m) long

Red Force,
Ferrari Land, Spain
112 mph (180 kph)
367 ft (112 m) high
2,890 ft (880 m) long

Kingda Ka
Six Flags Great Adventure,
New Jersey
128 mph (206 kph)
456 ft (139 m) high
3,118 ft (950 m) long

Ultimate
Lightwater Valley, UK
50 mph (80 kph)
107 ft (33 m) high
7,442 ft (2,268 m) long

Kingda Ka

This ride goes from
0–128 mph (206 kph) in
3.5 seconds, catapulting
riders as high as a
45-story building.

Montezum
Hopi Hari, Brazil
64 mph (103 kph)
139 ft (42 m) high
3,380 ft
(1,030 m) long

Colossos
Heide-Park, Soltau,
Germany
75 mph (102 kph)
197 ft (60 m) high
4,409 ft (1,344 m) long

ROLLER COASTERS AROUND THE WORLD
Numbers indicate ranking from 1-5.

Fastest	Highest	Longest	Unranked

Tower of Terror
Gold Reef City,
South Africa
59 mph (95 kph)
112 ft (34 m) high
328 ft (100 m) long

Roller coasters

Breakneck speeds, hair-raising twists and
turns, stomach-churning drops—roller
coasters can satisfy even hardened thrill-
seekers. This map shows some of the
world's biggest and best coasters.

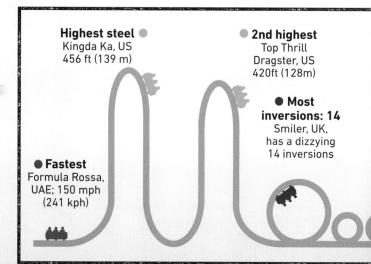

Highest steel
Kingda Ka, US
456 ft (139 m)

2nd highest
Top Thrill
Dragster, US
420ft (128m)

Most inversions: 14
Smiler, UK,
has a dizzying
14 inversions

Fastest
Formula Rossa,
UAE; 150 mph
(241 kph)

Flying roller coasters

These coasters—such as Manta at SeaWorld in Florida (right)—make you feel as though you are flying. The cars run on the underside of the track. Riders start in a seated position, but as the ride starts they are rotated to face the ground.

Steel Dragon 2000
Nagashima Spa Land, Japan
95 mph (153 kph)
318 ft (97 m) high
8,133 ft (2,437 m) long

● **Formula Rossa**
Ferrari World, UAE
150 mph (241 kph)
171 ft (52 m) high
6,791 ft (2,070 m) long

Do-Dodonpa
Fuji-Q Highland, Japan
107 mph (172 kph); 171 ft (52 m) high;
3,901 ft (1,189 m) long

Dinoconda
China Dinosaurs Park, China
80 mph (128 kph); 249 ft (76 m) high
3,471 ft (1,058 m) long

Ten Inversion Roller Coaster
Chimelong Paradise, China
45 mph (72 kph); 100 ft (30 m)
high 2,789 ft (850 m) long

Fujiyama
Fuji-Q Highland, Japan
81 mph (130 kph)
260 ft (70 m) high
6,709 ft (2,045 m) long

Takabisha
Fuji-Q Highland, Japan
62 mph (100 kph)
141 ft (43 m) high
3,281 ft (1,000 m) long

18 MPH
(29 KPH): SPEED OF THE WORLD'S OLDEST COASTER,
LEAP THE DIPS

4-D roller coasters

Fourth-dimension (4-D) coasters, such as China's Dinoconda, give theme parks an extra level of thrills. The seats on a 4-D coaster can rotate forward or backward, so as the riders hurtle along the track they also spin in a full circle. Eejanaika (below) is a 4-D ride at Japan's Fuji-Q Highland theme park.

DC Rivals Hypercoaster
Warner Bros. Movie World, Queensland, Australia; 71.5 mph (115 kph); 4,593 ft (1,400 m) long; 202 ft (61.6 m) high

Roller coaster records

Opened in 1902, the world's oldest coaster is the wooden Leap-the-Dips, at Lakemont Park, Pennsylvania. Since then, coasters have become taller, longer, faster—and scarier! Today's coasters are usually made of steel. Wood is less flexible than steel, so wooden coasters tend to be less complex and extreme than steel ones.

● **Steepest drop**
TMNT Shellraiser, US 121.5 degrees

● **Highest G-force**
Tower of Terror, South Africa
6.3G

National flags

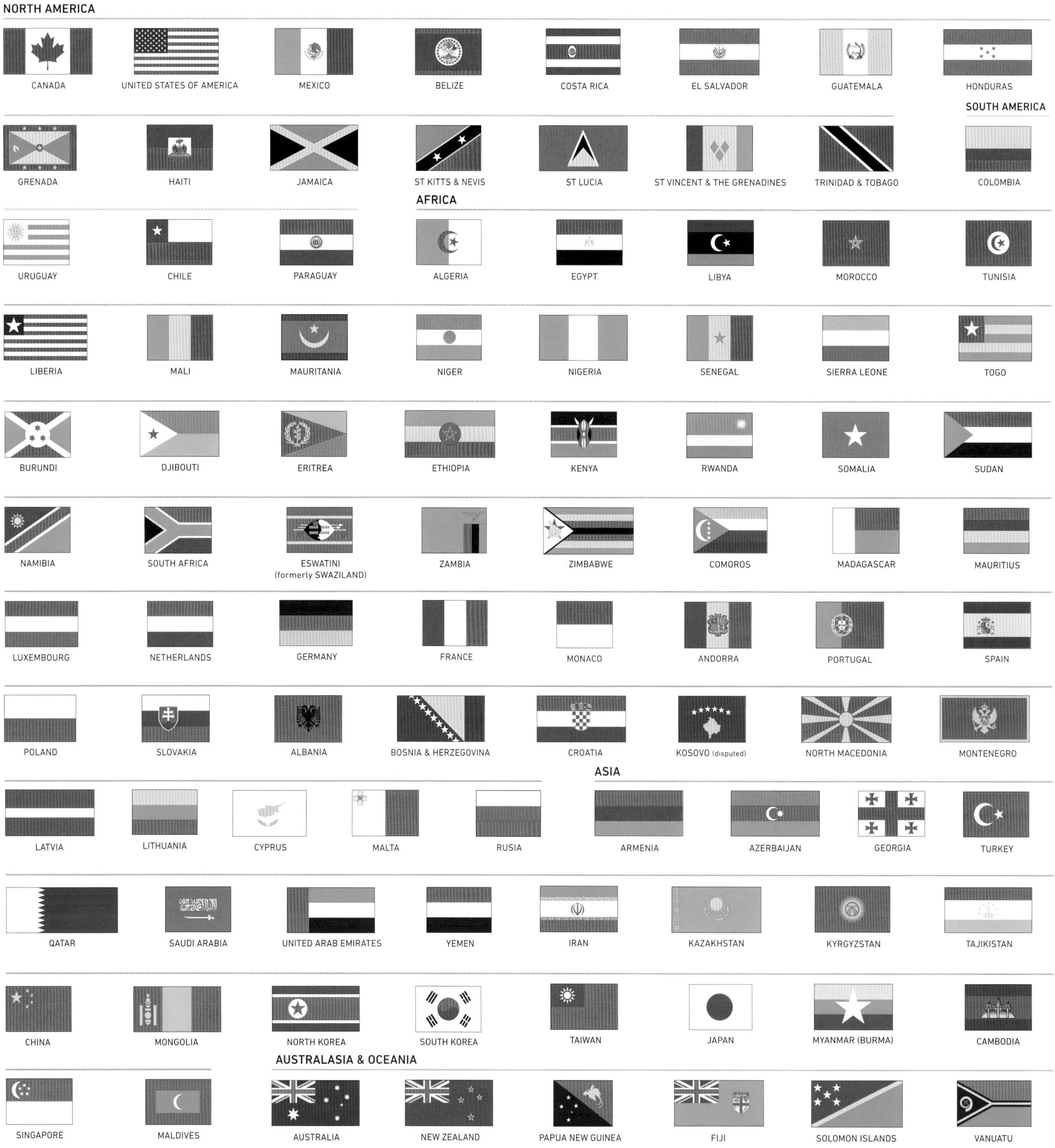
NORTH AMERICA
CANADA
UNITED STATES OF AMERICA
MEXICO
BELIZE
COSTA RICA
EL SALVADOR
GUATEMALA
HONDURAS
GRENADA
HAITI
JAMAICA
ST KITTS & NEVIS
ST LUCIA
ST VINCENT & THE GRENADINES
TRINIDAD & TOBAGO
SOUTH AMERICA
COLOMBIA
URUGUAY
CHILE
PARAGUAY
AFRICA
ALGERIA
EGYPT
LIBYA
MOROCCO
TUNISIA
LIBERIA
MALI
MAURITANIA
NIGER
NIGERIA
SENEGAL
SIERRA LEONE
TOGO
BURUNDI
DJIBOUTI
ERITREA
ETHIOPIA
KENYA
RWANDA
SOMALIA
SUDAN
NAMIBIA
SOUTH AFRICA
ESWATINI (formerly SWAZILAND)
ZAMBIA
ZIMBABWE
COMOROS
MADAGASCAR
MAURITIUS
LUXEMBOURG
NETHERLANDS
GERMANY
FRANCE
MONACO
ANDORRA
PORTUGAL
SPAIN
POLAND
SLOVAKIA
ALBANIA
BOSNIA & HERZEGOVINA
CROATIA
KOSOVO (disputed)
NORTH MACEDONIA
MONTENEGRO
LATVIA
LITHUANIA
CYPRUS
MALTA
RUSIA
ASIA
ARMENIA
AZERBAIJAN
GEORGIA
TURKEY
QATAR
SAUDI ARABIA
UNITED ARAB EMIRATES
YEMEN
IRAN
KAZAKHSTAN
KYRGYZSTAN
TAJIKISTAN
CHINA
MONGOLIA
NORTH KOREA
SOUTH KOREA
TAIWAN
JAPAN
MYANMAR (BURMA)
CAMBODIA
SINGAPORE
MALDIVES
AUSTRALASIA & OCEANIA
AUSTRALIA
NEW ZEALAND
PAPUA NEW GUINEA
FIJI
SOLOMON ISLANDS
VANUATU

A SOVEREIGN STATE IS A COUNTRY INDEPENDENT OF OTHER STATES, AND

OF ALL THE FLAGS OF THE WORLD'S **195 SOVEREIGN STATES,** ONLY **NEPAL'S** HAS MORE THAN **FOUR SIDES**

NICARAGUA · PANAMA · ANTIGUA & BARBUDA · THE BAHAMAS · BARBADOS · CUBA · DOMINICA · DOMINICAN REPUBLIC

GUYANA · SURINAME · VENEZUELA · BOLIVIA · ECUADOR · PERU · BRAZIL · ARGENTINA

BENIN · BURKINA FASO · CAPE VERDE · THE GAMBIA · GHANA · GUINEA · GUINEA–BISSAU · IVORY COAST

CAMEROON · CENTRAL AFRICAN REPUBLIC · CHAD · CONGO · DEM. REP. CONGO · EQUATORIAL GUINEA · GABON · SÃO TOMÉ & PRÍNCIPE

SOUTH SUDAN · TANZANIA · UGANDA · ANGOLA · BOTSWANA · LESOTHO · MALAWI · MOZAMBIQUE

EUROPE

SEYCHELLES · DENMARK · FINLAND · ICELAND · NORWAY · SWEDEN · IRELAND · UNITED KINGDOM · BELGIUM

ITALY · SAN MARINO · VATICAN CITY · AUSTRIA · LIECHTENSTEIN · SLOVENIA · SWITZERLAND · CZECHIA · HUNGARY

SERBIA · BULGARIA · GREECE · MOLDOVA · ROMANIA · UKRAINE · BELARUS · ESTONIA

IRAQ · ISRAEL · JORDAN · LEBANON · SYRIA · BAHRAIN · KUWAIT · OMAN

TURKMENISTAN · UZBEKISTAN · AFGHANISTAN · PAKISTAN · BANGLADESH · BHUTAN · INDIA · NEPAL · SRI LANKA

LAOS · PHILIPPINES · THAILAND · VIETNAM · BRUNEI · INDONESIA · EAST TIMOR · MALAYSIA

MARSHALL ISLANDS · MICRONESIA · NAURU · PALAU · KIRIBATI · TUVALU · TONGA · SAMOA

Index

A

Abu-Simbel 143
Abyssal plains 16
acid rain 99
Acropolis 143
adaptations 42–43
Afghanistan 83, 97, 142, 143, 155
Africa 26, 78, 80, 86, 89, 90, 94, 155
age profile 80–81
agriculture 75, 92–93, 102
air pollution 98–99
air travel 85, 116–17, 160–61
aircraft, military 130–31
airports, busiest 116
Aksum stelae 142–43
Alaska 10, 14, 32, 40–41, 54
Aleutian Trench 9, 16
Alexander the Great 135, 141
algae 64
Algeria 24
alternative energy 74, 106–07
aluminum 100–01
Amazon Rainforest 32, 64, 110
Amazon, River 20, 21, 56
American Civil War 135, 152
Amoco Cadiz 158–59
Amur-Arqun 20, 21
ancient civilizations 140–43, 152–53
Andes 12, 24, 66–67
animals see wildlife
Antarctica 7, 26–27, 34–35, 36–37, 55
Antioch 11
ants 60
apartheid, end of 135
Arab Spring 134, 157
arachnids 48–49, 64
arapaimas 58
architecture
    castles 150–51
    medieval 146–47
    modern era 160–61
    tallest buildings 124–25
Arctic 7, 31, 36, 64, 65, 74, 75
Arctic terns 52–53
Argentina 13, 44, 54, 86, 106, 178
USS Arizona 159
armed forces 130–31, 152–53
art 164, 165, 172–75
    prehistoric 135, 138–39
Artemis, Temple of 143
Ashoka, Emperor 152–53
asteroid impact 10, 22
Atacama Desert 34
Atlanta 116
atmosphere 6, 104, 108
Australia
    culture 167, 173, 177, 179, 181
    land 22, 24, 27, 29, 33
    living world 45, 67
    people 77, 83, 89, 92, 95, 103, 107

Australopithecus 136–37
Austria 101
autobahns 115, 122
Aztec Empire 135, 146, 148, 149

B

Baikal, Lake 21
Bali 165
Bamiyan Buddhas 142, 143
Bangladesh 26, 27, 29, 77
Barringer Crater 23
Basawan 173
basins, oceanic 16
Batavia 159
battlegrounds 152–53
beaches 170–71
bees 48, 60, 61
beetles 60, 61
Beijing 76, 77, 116, 117, 151, 181
Belarus 32, 83
Belgium 179
Bester, Willie 173
Bettencourt Meyers, Françoise 90
Bezos, Jeff 91
Bhola Cyclone 29
Bhutan 83
big wheels 170–71
billionaires 90–91
biodiversity 64–65
biofuel, biogas, and biomass 106–07
bioluminescence 42
biomes 30–31, 67
biosphere 7, 74
Bird Flu 85
birds 42, 46–53, 68–71
HMS Birkenhead 159
Bismarck 159
Black Death 84, 85
blue whales 54–55
bog bodies 144
Bolivar, Simón 156
Bolivia 82, 86, 94, 135, 156
boreal forests 30, 33
Borneo 27, 33, 65
Borobudar, Java 147
boundaries, plate 8–9
boxing 181
Brazil 10, 26, 54, 76, 92, 96, 103,
    106, 107, 130, 172, 176, 180,
    181, 183
bridges 115, 120, 123, 135, 161
Britain, Battle of 152
British Empire 135, 154–55
broadband 127
Brooklyn Bridge 115
bubonic plague 84, 85
Buddhism 168, 169
Burghausen 150
burial sites 135, 139, 144–45

Burj Khalifa 112–13, 124, 125
Burundi 87
butterflies 60, 61, 69, 70
Byzantine Empire 135, 149

C

Cajamarca, Battle of 135, 152
California 32, 50, 66
calories, daily intake of 94–95
Cairo 76
Cambodia 103
Cameroon 95, 166
Canada
    culture 177, 179, 180
    land 22, 24
    living world 44
    people 80, 88, 92, 94, 96, 98, 104,
        106, 107, 110
Canary Islands 66
Cape Town 117
Caravaggio 173
carbon dioxide 99, 108
cargo 118–19
Carnival 176–77
carnivorous plants 60–61
Carthage, Siege of 153
Castle of Good Hope 151
castles 150–51
Central African Republic 97
Cerro el Cóndor 13
Chad 96
Chagall, Marc 173
Channel Tunnel 160–61
channels, TV 178
chemical pollution 98–99
Chesapeake Bay 23
Chicago 116
Chicxulub 23
Chile 10, 12, 13, 92, 145
Chimborazo, Mount 12
Chimu Empire 148, 149
China
    armed forces 131
    culture 167, 169, 173, 175, 177, 178,
        179
    history 134, 137, 142, 143, 151, 156
    land 11, 12, 25, 26
    living world 44–45, 67
    people 77, 81, 87, 89, 93, 95, 97, 99,
        101, 105, 107
Chinese New Year 177
Christianity 148, 168–69
Chrysler Building 124
cicadas 60, 61
cities, biggest 76–77
civilizations 134, 140–41, 148–49
climate change 98, 108–09
clothing 164
clouds 6

coal 104–05, 106, 107
coffee 92–93
cold deserts 35
Colombia 15, 135, 156
colonialism 154–55
Colosseum 135, 142–43
Colossus of Rhodes 143
Columbus, Christopher 146
Communism, collapse of 135, 157
computer technology 114, 126–27
Concorde 134, 161
concrete 115
Congo-Chambeshi 20
conservation 75, 110–11
Constantinople, Fall of 135, 153
construction 115, 124–25
continental crust 9
continental shelf 17
convection currents 7
convergent boundaries 8
Coordinated Universal Time (UTC) 38
coral/coral reefs 30, 42, 111
Coral Sea, Battle of the 152, 153
core, Earth's 6, 7
cost of living 86–87
Costa Rica 130
COVID-19 85, 86, 134
craters 22–23
Crécy, Battle of 152
Cretaceous Period 44–45
cricket 181
crocodiles 49, 58–59
crops 92–3
Crusades 134, 153
crust, Earth's 6, 7, 8–9
crustaceans 64
Cuba 66, 83, 94, 152, 156
culture 162–87
    prehistoric 138–39
currencies 89
currents, ocean 18–19
cycling 181
cyclones 28–29

D

Dallas 116
dance 164–65
Dangote, Aliko 91
Darfur 26
day and night 38
deep water currents 19
deforestation 32–33
Delacroix, Eugène 172
Delhi 76, 77, 117
Democratic Republic of Congo 106
Denmark 166, 179
deserts 4–5, 24, 31, 34–35
    life in 42–43, 64
    nomads 78, 79

Dhaka 76, 77
dinosaurs 10, 22, 44–45
divergent boundaries 8
diving and snorkelling 70–71
Diwali 165
doctors, per capita 83
Dominican Republic 26
MV Doña Paz 159
dragonflies 60
drones, unmanned 130
droughts 103
Dubai 112–13
dunes 35

E

Earnhardt, Dale Snr. 182
Earth
  interior of 6
  rotation of 7, 38
  structure of 6–7
earthquakes 8, 10–11
East African Rift 8, 15
East Melanesia 67
East Pacific Rise 9, 16
Easter Island 132–33, 174, 176
Ecuador 12, 135, 156
education 96–97
Egypt 24, 53, 92, 130, 131
  ancient 134, 135, 140, 143, 144–45
El Salvador 106
Emperor Seamounts 17
Empire State Building 125
empires
  ancient 140–41
  colonial 154–55
  medieval 148–49
endemic hot spots 67
energy
  alternative 74, 106–07
  resources and consumption 74,
    104–05
ENIAC 114
Eritrea 95
erosion 20
Eswatini 82–83, 177
Ethiopia 67, 155
  empire 148
Europe, literacy in 96
Everest, Mount 12, 13, 16
extinctions 10, 22, 50–51, 68, 69, 70–71

F

Falkland Islands 104, 178
fashion 164
fault lines 9
festivals 162–63, 165, 176–77
Finland 177, 179
fish 46, 47
  dangerous 48–49
  river 58–59
fishing industry 92, 93
flags 186–87
flash floods 26
fleas 50, 61

flooded savanna 30
floods 26
floral kingdoms 62
flu viruses 84–85
food
  cookery 164
  cost of 95
  intake 94–95
  production 92–93
  supplies 82
food chains 47
football (soccer) 180–81
footprint, human 74–75
footprints, dinosaur 45
Forbidden City, Beijing 151
forests 30, 32–33, 110–11
Formula 1 (F1) 182–83
Fort Independence, Boston 151
fossil fuels 74, 104–05, 106, 107
fossils 44–45, 136–37
France 89, 92, 104, 106, 122, 130, 131,
  154–55, 172, 178, 179
Frankfurt 116
freeways 115, 122
French Guiana 178
French Revolution 135
freshwater creatures 56, 58–59
Fukuoka 117
fungi 64

G

Gabon 100
Gainsborough, Thomas 172
Galápagos Islands 50, 66
Gandhi, Mahatma 156
Gansu earthquake 11
garbage patches 19, 100–01
gas 104–05, 106, 107
Gates, Bill 90
gender differences 97
Genghis Khan 134, 149, 175
Georgia 83, 97
geosynchronous orbit 128, 129
geothermal energy 106–07
Germany 44, 89, 101, 106, 107, 115,
  136, 151, 154–55, 178, 179, 180, 183
Ghana 86, 88, 156, 178
  ancient 148
giant catfish 58–59
Gibraltar 53
glaciers 37, 108–09, 110
global warming 98, 108–09
gold 88–89
GPS satellites 129
Graf Zeppelin airship 161
grasslands 30, 35
Great Dying 10, 22
Great Game 155
Great Lakes 20
Great Sphinx 134, 174–75
Great Stupa of Sanchi 142
Great Wall of China 135, 142, 143
great white sharks 48, 56–57
Great Zimbabwe 134, 148, 151
Greeks, ancient 142, 143, 153

Greenland 24, 53, 80, 110
  ice sheet 34, 109
Greenwich Mean Time (GMT) 39
Guatemala 14, 80, 95
Guevara, Che 156
Guinea-Bissau 82, 97
Gulf Stream 19
Gulf War 98
Guyana 80, 103
gyres 18, 19, 100

H

habitats
  and adaptations 42–43
  destruction of 68–69
  unusual 66–67
Hagia Sofia 143
Haiti 11, 26, 86, 102
Halincarnassus, Mausoleum at 143
Hamilton, Lewis 183
Han Empire 135, 141
Hanging Gardens of Babylon 143
Harvey, Hurricane 28
Hawaii 13, 14, 28, 38, 66
health 82–85, 98–99
Himalayas 8, 13, 65, 109
Himeji 151
Hinduism 168, 169
history 132–61, 174–75
HIV/AIDS 85
Hokusai, Katsushika 173
Holi Festival 162–63
Holy Roman Empire 135, 149, 152
Homo genus 134, 136–37
Hong Kong 116, 117, 127
Hong Kong Flu 85
Hoover Dam 161
Hopper, Edward 172
Huari Empire 135, 148, 149
Hubble Space Telescope 129
humans
  early 136–37
  impact of 74–75
hurricanes 28–29
hydroelectric energy 106–07

I

ice 7, 36–37
ice sheets 36, 37, 108, 110
icebergs 37, 158
Iceland 14, 16, 77, 87, 106–07, 166
Idai, Cyclone 29
impact craters 22–23
Inca Empire 148, 149
income, per capita 86–87
India
  armed forces 131
  culture 162–63, 164–65, 167, 173,
    177, 178, 181
  history 134, 142, 151, 152–53, 157
  land 12, 27, 39
  people 77, 81, 87, 89, 93, 95, 99,
    103, 107
Indian Ocean 10

indigenous peoples 78–79, 111
Indo-Pakistani War 134, 153
Indochina War, First 134, 153
Indonesia 14, 15, 89, 97, 99, 103, 107,
  137
Industrial Revolution 160
industrial waste/accidents 98–99
industrial wonders 160–61
inequality 86–87
infectious diseases 84–85
information technology 126–27
infrastructure 115, 120–23
Iniki, Hurricane 28
insects 48–51, 60–61, 64
International Date Line 38
International Monetary Fund 89
International Space Station 129
International Union for Conservation
  (IUNC) 68
Internet connections 126–27, 164
Inuit 75, 78
invasive species 50–51
invertebrates 64
Iran 26, 131, 181
Iraq 25, 103
Ireland 95, 179
Islam 148, 168, 169
Israel 25, 130, 131
Italy 89, 92, 106, 154–55, 173, 179

J

Japan
  culture 169, 173, 177, 179
  history 145, 151, 154–55
  land 10, 15, 27, 29, 33
  people 77, 81, 83, 89, 92, 93, 99, 107
Jeju 117
Jerusalem 153, 168
jewelry, first 135, 138
Johannesburg 117, 181
Juanita the Ice Maiden 145
Judaism 168
Jurassic Period 44

K

K2 12
Kahlo, Frida 172
kakapo (owl parrot) 68–69
Kalinga, Battle of 152–53
Kamchatka earthquake 10
Kanem Empire 134, 149
Kangchenjunga 12
Kathakali dancers 164–65
Katrina, Hurricane 28, 29
Kazakhstan 103
Kenya 92, 95, 103, 107
Khmer Empire 149
Kiribati 38
Kolkata 77
Korean War 134, 153
Krak des Chevaliers 150
Krakatau 14, 15
Kuwait 25, 101, 103
Kyrgyzstan 103

L

Lalibela 147
lakes 6, 20–21, 109
land ice 36
landfill 100, 101
languages 164, 166–67
Large Hadron Collider 160
Le Mans 183
lead pollution 98–99
Leaning Tower of Pisa 147
Lempicka, Tamara de 173
Lenin, Vladimir 156
Lhotse 12
Liberia 83, 86, 155
Liberty, Statue of 174–75
Libya 24
lichens 64
Liechtenstein 130
life on Earth 6, 7, 40–71
life expectancy 82–83
Lindow Man 144
literacy 96–97
literature 165
livestock 92–93
Llullaillaco 13
locusts 60
London 116
Los Angeles 72–73, 116
Low Earth Orbit (LEO) 129
RMS Lusitania 159
Luxor 24

M

Macedonian Empire 135, 141
Machu Picchu 135, 146
Madagascar 67, 97
Makalu 12
Malawi 83
Malaysia 81, 175, 179
Mali 134, 148, 176
malnutrition 94–95
Malta 53
mammals 46–51, 68–71
mangrove 30
Manila 77
mantle 6, 7
Mao Zedung 134, 156
Marble Bar 24
Marcus, Cyclone 29
Mariana Trench 17
marine animals 42, 48–49, 54–57
marine biomes 30
Mars 12–13
Martinique 15
Mauna Kea 13
Mauritania 96, 102
Mauryan Empire 135, 141, 153
Mayan civilization 135, 140, 141, 146
mayflies 60
median age 80–81
medical care 82, 83
medieval age 146–49, 152–53
Mehrangarh Fort, Jodhpur 151
Meirelles, Victor 172
Melbourne 117

mercury, toxic 99
meteorites 22–23
Mexico 24, 28, 54, 66, 76, 80, 98, 106, 142, 144, 172, 175, 176, 180
Mexico City 76, 180
Mid-Atlantic Ridge 8, 14, 16
mid-ocean ridges 16–17
Middle East, oil 105
midges 60, 61
migration
    animals 170–71
    birds 52–53
    human 78–79, 164
    insects 60, 61
    sharks 48, 49
    whales 55
military forces 130–31
minerals 74
mines, gold 88
Ming Dynasty 135, 149
Mississippi–Missouri 20, 26, 56
mollusks 64
Monaco 82, 83
monarch butterflies 60, 61
Monet, Claude 172
Mongol Empire 134, 149
Mongolia 45, 77, 95, 175, 178
monsoon 27
Morocco 80, 86
mosquitoes 60
moths 60
motor racing 182–83
mountains 6, 12–13, 16–17, 122
Mozambique 83, 97
Mughal Empire 135, 149
Mumbai 76, 77, 117
mummies 144–45
Munch, Edvard 172–73
music 135, 138, 164, 165
Musk, Elon 91
Myanmar (Burma) 175

N

Namib Desert 4–5, 34
Namibia 4–5, 77, 87
NASCAR sites 182
national parks 110–11
native species 50–51
natural resources 74, 102–05
Nauru 94
Neanderthals 136, 137
Nepal 12, 175, 187
Netherlands 89, 92, 101, 154–55, 179
Nevado de Incahusai 13
Nevado del Ruiz 15
Nevados Ojos de Salado 13
New Caledonia 67
New York City 76, 115, 174
New Zealand 27, 33, 55, 81, 93, 97, 177
nickel 99
Niger 83
Nigeria 100, 104, 172
night and day 38
Nile, River 20

Nkrumah, Kwame 156
nomads 78–79
Norte Chico civilization 135
North Korea 131, 174, 181
North Sea 104
Norway 39, 87, 101, 102, 106, 107, 166, 172, 179
Novarupta 14
nuclear energy 106–07
nuclear waste/accidents 98–99
nuclear weapons 130–31
Nuestra Señora de Atocha 159

O

Ob-Irtysh 20–21
obesity 94
ocean floor 6, 16–17
oceanic crust 9
oceans 7
    and climate change 108–09
    conservation 110–11
    currents 18–19, 24–25
    life in 42, 47, 48–49, 54–57
    pollution 19, 98, 100–01
oil
    resources 104–05, 106, 109
    spills 98, 158–59
Olduvai Gorge 136
Olmec civilization 135, 140, 141
Olympus Mons 12–13
Oman 179
Ortega Gaona, Amancio 91
Osaka 76, 117
Ottoman Empire 149, 152–55
Ötzi the Iceman 144

P

Pacific Ring of Fire 14, 15
paintings 139, 165, 172–73
Pakistan 12, 25, 92, 95, 131
Palermo 145
Panama 53
Panama Canal 118, 160
pandemics 84–85
Papua New Guinea 33, 67, 81, 97, 167
Paraguay 166
Paraná 20
Paranthropus 136–37
parasites 50
Paris 27, 116
passengers, air 116–17
passes, mountain 122
Patagonian Desert 34
Patricia, Hurricane 28
Pelée, Mont 15
peregrine falcons 46–47
Persian Empire, First 135, 141
Persian Gulf 98
Peru 88, 92, 98, 102, 135, 142, 145, 152, 156, 166, 176
Peru-Chile Trench 9, 16
pesticides 98–99
pests 50–51
Petra 143

Petronas Towers 124, 125
Pharos of Alexandria 143
Philippines 14, 67, 77, 107, 144
Picasso, Pablo 172, 173
Pinatubo, Mount 14
plague 84–85
plants 6, 7, 62–63
    adaptations 42–43
    biodiversity 64–65
    biomes 30–31
    invasive species 50–51
    unique 66–67
plastic waste 100–01
plate tectonics see tectonic plates
poison-dart frogs 48–49, 65
Poland 32, 173, 174
polar regions 7, 36–37
    deserts 31, 35
    life in 43
pollution 75, 98–99, 104, 108
Polynesia 66
Pont-du-Gard 142, 143
pop music 164, 180
population
    age profile 80–81
    distribution 76–77, 110–11
    and food supplies 93
    growth 74–75
ports, busiest 119
Portugal 154, 181
pottery 139
poverty 86–87
predators 46–47
prehistory 136–39
Prime Meridian 39
Prince William Sound 10
Puffing Billy 115
pyramids 142–43, 146

R

radioactive waste 98–99
railroads 114–15, 120–21, 160
rainfall 5, 6, 26–27
rainforests 32–33, 43, 64, 65
Ramayana 164–65
rats 50
recycling 74, 100–01, 103
Red List (IUCN) 60
religion 168–69, 175, 176–77
renewable energy 74, 106–07
reptiles 43, 46–51, 58–59
Réunion 27, 29
revolutions 152–53, 156–57
rice production 93
Rio de Janeiro 26, 117, 176, 180
Rio de la Plata 20
rivers 6, 20–21
river monsters 58–59
roads 115, 122–23
Rocky Mountains 12
roller coasters 184–85
Romania 178
Romans 115, 135, 141, 153
rubbish 100–01

Russia
  armed forces 131
  culture 167, 173, 174, 175, 177
  history 135, 154–55, 156, 157
  land 10, 24, 25, 26, 39
  people 87, 89, 91, 92, 97, 99, 103, 105, 107
Ruwenzori Mountains 13

S

safaris 170–71
Sahara Desert 34–35, 64, 110
St. Peter's Basilica, Rome 147
salt 19
San Andreas Fault 9
Santa Maria volcano 14
São Paulo 76, 117
Sapporo 117
satellites 128–29
Saudi Arabia 94, 105, 131
savanna 30
Schumacher, Michael 183
Scramble for Africa 155
sculpture 139, 165, 172, 174–75
sea ice 36, 109
sea levels 108–09
sea transportation 118–19
seamounts 16–17
secondary education 96
seismic waves 10
semideserts 35
Senegal 26, 174
Senna, Ayrton 183
Seoul 117
Seven Wonders of the World 142–43
Shaanxi earthquake 11
Shanghai 76, 77, 117, 119
sharks 46, 47, 48, 56–57
sheep 93
Shinto 168, 169
shipping routes 118–19
shipwrecks 158–59
shrubland 31
Sicily 53, 145
sieges 153
Sikhism 168, 169
Singapore 24
skyscrapers 112–13, 115, 124–25, 160
slave trade 155
snakes 43, 46–51
snow 6, 26–27
solar energy 74, 106–07
Solomon Islands 101
Somalia 25, 97
Somme, Battle of the 134–35, 152
Songhai Empire 149
South Africa 55, 67, 87, 89, 99, 136, 151, 173, 177, 178, 179, 181
South Korea 101, 131, 177, 181
South Sudan 26, 83, 123,
space debris 128–29
Space Shuttle 128
Spain 106, 107, 154, 166, 172, 176, 177, 180, 183

Spanish flu 84, 85
speedway 182
sperm whales 55
spiders 48–49
sport 180–83
Sri Lanka 55, 67, 177
stadiums 164, 180–81
statues 174–75
steam engines 115
Stone Age 138–39
Stonehenge 142, 143
streaming 179
submarines 130–31
Sudbury Basin 23
Suez Canal 119
sun, energy from 7
Sundaland 67
superbugs 85
surface currents 18
Suriname 77, 103, 166
HMS Sussex 159
swarms 60–61
Sweden 24, 83, 87, 101, 107, 166, 179
Swine Flu 85
Switzerland 89, 99, 100–01, 179
Sydney 117
Sydney Opera House 135, 161
Syria 151

T

Taipei 101, 117, 124, 125
Tajikistan 103
Tambora 14, 15
Tangshan earthquake 11
tanks, battle 130–31
Tanzania 25, 136
tea trade 92
tectonic plates 8–9, 10, 12, 14, 16, 17
telecommunications 115, 126–27, 160
television 178–79
temperate biomes 30, 32
temperatures 24–25, 108–09
termites 60
Terracotta Army 142, 143
Thailand 107, 181
Thanksgiving 176–77
time zones 38–39
Tip, Typhoon 29
Tipas 13
RMS Titanic 135, 159
Tiwanaku Empire 148, 149
Tohoku earthquake 10
Tokyo 76, 77, 116, 117
Tonga 94
tools, early 134, 138
tourism 170–71
towers, unsupported 125
trade 118–19
trains 114–15, 120
transform boundaries 8
transportation 114–23
trenches, ocean 8, 9, 16–17
Triassic Period 44
Trinidad and Tobago 98, 104
tropical cyclones 28–29

tropical forests 30, 33, 64, 65
tsunamis 8
tundra 31, 35, 78, 110
Tunisia 24, 176
tunnels, longest rail 121
Turkey 11, 178, 181
Turkmenistan 103
Tutankhamun 144

U

Uganda 81
Ukraine 98, 107, 174
Umayyad Caliphate 135, 149
United Arab Emirates 94, 112–13
United Kingdom
  armed forces 130, 131
  culture 172, 174–75, 176, 178, 179, 180, 181
  history 135, 152, 154–55
  people 92, 94, 95
  time zone 38
United States
  armed forces 130, 131
  culture 166, 172, 176, 178–79, 180, 182
  history 151, 152, 153, 158, 160
  land 23, 24, 26, 28, 38
  living world 44–45
  people 76, 80, 86, 88, 89, 91, 92, 94, 95, 96, 98, 99, 101, 102, 103, 104, 105, 106, 107
Unzen, Mount 15
Uruguay 80
USSR 157
Uzbekistan 103

V

Valdivia earthquake 10
Vanuatu 167
vegetation
  biomes 30–31
  deserts 34–35
  forests 32–33
  wilderness 110–11
Velaro 114–15
Venezuela 104, 106, 135, 156, 175
venom
  animals 48–49, 65
  plants 62–63
Verkhoyansk 24, 25
vertebrates 64
Very Large Array 135, 160
Victoria, Lake 21
Vienna, Battle of 135, 152
Vietnam 87, 93, 134
viruses 84–85
volcanoes 8, 13, 14–15
Vredefort impact structure 23

W

Wallacea 67
warfare 130–31, 152–53
Warhol, Andy 172
warships 130–31

wasps 60
waste 100–01
water
  clean 82, 102–03
  human consumption 102, 103
  pollution 98–99
  use of 75, 102
water cycle 6
Watson, Yannima Tommy 173
wealth 75, 86–91
weapons 130–31
weather 6
weevils 64
weight 94–95
Welwitschia 60–61
whales 40–41, 46, 47, 54–55
wheat 92
wilderness 100–11
wildlife
  adaptations 42–43
  biodiversity 64–65
  conservation 110–11
  deadly 48–49
  deserts 34–35
  endangered 66, 68–69
  extinct 44–45
  invasive species 50–51
  marine 42, 48–49, 54–57
  predators 46–47
  unique 66–67
  see also specific types
Wilhelm Gustloff 159
Willis Tower 124, 125
wind energy 74, 106–07
Windsor Castle 151
Winston, Cyclone 29
world parties 177
World War I 134–35, 152, 153, 158
World War II 134, 152, 153, 159
wrestling 181

Y

Yangtze River 20, 21, 26
Yellow River 20, 21
Yemen 97
Yenisei-Angara-Selenga 20, 21
Yue Minjun 173

Z

Zambia 99
Zeus, statue in Olympia 143
Zhoukoudian Caves 137
Zhucheng 44, 45
Zimbabwe 134, 148, 151
Zuckerberg, Mark 91

Acknowledgments

Dorling Kindersley would like to thank: Caitlin Doyle for proofreading, Helen Peters for indexing, Haisam Hussein, Anders Kjellberg, Peter Minister, Martin Sanders, and Surya Sarangi for illustration, Deeksha Miglani and Surbhi N. Kapoor for research, and David Roberts for cartographic assistance.

The publisher would like to thank the following for their kind permission to reproduce their photographs:

(Key: a-above; b-below/bottom; c-center; f-far; l-left; r-right; t-top)

2 Andy Biggs: www.andybiggs.com (tc). **Corbis:** Alaska Stock (tr). **3 Corbis:** Floris Leeuwenberg (ftr); SOPA / Pietro Canali (tl). **Getty Images:** Art Wolfe (tr). **Sebastian Opitz:** (tc). **4–5 Andy Biggs:** www.andybiggs.com. **22 Getty Images:** Mark Garlick (br). **23 Corbis:** Charles & Josette Lenars (cr). **24–25 Robert J. Hijmans:** Hijmans, R.J, S.E. Cameron, J.L. Parra, P.G. Jones and A. Jarvis, 2005. Very high resolution interpolated climate surfaces for global land areas. International Journal of Climatology 25: 1965–1978 (base-map data). **26–27 Robert J. Hijmans:** Hijmans, R.J, S.E. Cameron, J.L. Parra, P.G. Jones and A. Jarvis, 2005. Very high resolution interpolated climate surfaces for global land areas. International Journal of Climatology 25: 1965–1978 (base-map data). **28–29 Adam Sparkes:** Data of the tropical cyclones projected by Adam Sparkes. Base image: NASA Goddard Space Flight Center Image by Reto Stöckli (land surface, shallow water, clouds). Enhancements by Robert Simmon (ocean color, compositing, 3D globes, animation). Data and technical support: MODIS Land Group; MODIS Science Data Support Team; MODIS Atmosphere Group; MODIS Ocean Group Additional data: USGS EROS Data Center (topography); USGS Terrestrial Remote Sensing Flagstaff Field Center (Antarctica); Defense Meteorological Satellite Program (city lights). **29 NOAA:** (tc). **30 Dorling Kindersley:** Rough Guides (tl, tr). **Shutterstock:** Edwin van Wier (crb). **31 Dreamstime.com:** (tc). **PunchStock:** Digital Vision / Peter Adams (tr). **35 NASA:** Goddard Space Flight Center, image courtesy the NASA Scientific Visualization Studio, (bl). **36 Dorling Kindersley:** Rough Guides / Tim Draper (bl). **Dreamstime.com:** Darryn Schneider (tr). **40–41 Corbis:** Alaska Stock. **42 Alamy Images:** Martin Strmiska (bl). Getty Images: Werner Van Steen (c). **43 NHPA / Photoshot:** Ken Griffiths (cr). **45 Corbis:** Science Faction / Louie Psihoyos (tr). Dorling Kindersley: Christian Williams (tc). **48 Alamy Images:**

National Geographic Image Collection (bl). **Dorling Kindersley:** Courtesy of the Weymouth Sea Life Centre (bc). **49 Dreamstime.com:** Francesco Pacienza (tr). **53 Corbis:** Roger Tidman (bc). **55 Corbis:** Paul Souders (ca). **56 Corbis:** Minden Pictures / Mike Parry (cl); National Geographic Society / Ben Horton (tc). **60 Dorling Kindersley:** Courtesy of the Natural History Museum, London (cra, c). **Getty Images:** Visuals Unlimited, Inc. / Alex Wild (cr). **61 Alamy Images:** Premaphotos (tl). **Corbis:** Visuals Unlimited / Robert & Jean Pollock (tr). **Getty Images:** Mint Images / Frans Lanting (tc). Photoshot: Gerald Cubitt (br). **62–63 Dreamstime.com:** Jezper. **62 Alamy Images:** Tim Gainey (bc); John Glover (br). FLPA: Imagebroker / Ulrich Doering (cb). **Getty Images:** Shanna Baker (clb); Alessandra Sarti (bl). **64 Dorling Kindersley:** Courtesy of Oxford University Museum of Natural History (clb). **64–65 Dr. Clinton N. Jenkins:** Data: IUCN Red List of Threatened Species / www.iucnredlist.org / BirdLife International; Processing: Clinton Jenkins / SavingSpecies.org; Design & Render; Félix Pharand–Deschênes / Globaia.org. **66 Dorling Kindersley:** Rough Guides (cl). **67 Corbis:** Ocean (crb). **Dorling Kindersley:** Roger and Liz Charlwood (crb/New Caledonia). **72–73 Corbis:** SOPA / Pietro Canali. **74–75 Getty Images:** Doug Allan. **75 Corbis:** Aurora Photos / Bridget Besaw (tl); Frank Lukasseck (ftl); Minden Pictures / Ch'ien Lee (tc); John Carnemolla (tr). **76–77 Center for International Earth Science Information Network (CIESIN):** Columbia University; International Food Policy Research Institute (IFPRI); The World Bank; and Centro Internacional de Agricultura Tropical (CIAT). **84 Corbis:** Dennis Kunkel Microscopy, Inc. / Visuals Unlimited (tc); Dr. Dennis Kunkel Microscopy / Visuals Unlimited (tr). **85 Dreamstime.com:** Lukas Gojda (cr). **89 Dreamstime.com:** Commeraydavo (tr). **90 Getty Images:** AFP / Martin Bureau (br). James Leynse (bc). **91 Corbis:** epa / Justin Lane (bl); Kim Kulish (cra); epa / Mario Guzman (br). **Getty Images:** AFP (cr); Bloomberg / Wei Leng Tay (bc). (bc). **93 Dreamstime.com:** Kheng Guan Toh (br). **101 Corbis:** Peter Adams (bl). **105 Corbis:** Shuli Hallak (bc). **107 Dreamstime.com:** Milosluz (bc). **108–109 NASA:** Goddard Space Flight Center Scientific Visualization Studio. **109 NASA:** 1941 photo taken by Ulysses William O. Field; 2004 photo taken by Bruce F. Molnia. Courtesy of the Glacier Photograph Collection, National Snow and Ice Data Center / World Data Center for Glaciology. (bl). **110–111 UNEP–WCMC:** Dataset derived using the Digital Chart of the World 1993 version and methods based

on the Australian National Wilderness Inventory (Lesslie, R. and Maslen, M. 1995. National Wilderness Inventory Handbook. 2nd edn, Australian Heritage Commission. Australian Government Publishing Service, Canberra) (base-map data). **112–113 Sebastian Opitz. 114–115 Dreamstime.com:** Dmitry Mizintsev (c). **114 Corbis:** (bc); Science Faction / Louie Psihoyos (br). **115 Corbis:** Bettmann (crb); Cameron Davidson (br). **Dorling Kindersley:** Courtesy of The Science Museum, London (tc). **Getty Images:** Three Lions (bc). **116–117 Michael Markieta:** www.spatialanalysis.ca. **118–119 Prof. Dr. Bernd Blasius:** Journal of the Royal Society Interface, The complex network of global cargo ship movements, p1094, 2010 (base-map data). **122 Getty Images:** Radius Images (bc). **126–127 Chris Harrison** (base-map). **128–129 ESA. 128 NASA: Columbia Accident Investigation Report, (bc). 129 ESA: (cra). NASA:** Image created by Reto Stockli with the help of Alan Nelson, under the leadership of Fritz Hasle (br). **130 Corbis:** DoD (br). **132–133 Getty Images:** Art Wolfe. **134 Corbis:** Radius Images (bl); **Getty Images:** (cr). **Dreamstime.com:** Kawee Srital On (cb). **135 Corbis:** Sodapix / Bernd Schuler (b). **136–137 Corbis:** W. Cody. **137 Science Photo Library:** MSF / Javier Trueba (crb). **138 akg-images:** Oronoz (clb/Mousterian Tool). **Dorling Kindersley:** The American Museum of Natural History (bl); Natural History Museum, London (cl, clb). **Getty Images:** AFP (tc); De Agostini (tr). **139 akg-images:** Ulmer Museum (bc). **Getty Images:** De Agostini (crb). **141 Dorling Kindersley:** Courtesy of the University Museum of Archaeology and Anthropology, Cambridge (tl); Ancient Art / Judith Miller (bc/Urn); Alan Hills and Barbara Winter / The Trustees of the British Museum (tc); Stephen Dodd / The Trustees of the British Museum (tr). **Getty Images:** De Agostini (bl). **144 Alamy Images:** Ancient Art & Architecture Collection Ltd (tc). **Getty Images:** Copper Age (tl). Rex Features: (tr). **148 Dorling Kindersley:** © The Board of Trustees of the Armouries (tr); The Wallace Collection, London (cb). **149 Dorling Kindersley:** © The Board of Trustees of the Armouries (cla); Lennox Gallery Ltd / Judith Miller (cra); William Jamieson Tribal Art / Judith Miller (bl); Courtesy of the Royal Armories (tc); The Trustees of the British Museum (cb); Peter Wilson / CONACULTA–INAH–MEX. Authorized reproduction by the Instituto Nacional de Antropología e Historia (clb). **150 Corbis:** Walter Geiersperger (cl); Robert Harding World Imagery / Michael Jenner (clb). **151 Alamy Images:** Peter Titmuss (bc). **Corbis:** Design Pics / Keith Levit (cra). **Dreamstime.com:**

(bl). Getty Images: AFP (cr). **156 Corbis:** Bettmann (cb, cra). **Getty Images:** (c). **157 Corbis:** Bryan Denton (bl); Peter Turnley (cr). Getty Images: AFP (ca); (c); (clb). **159 Dreamstime.com:** (bc). **162–163 Corbis:** Floris Leeuwenberg. **164 Getty Images:** Redferns / Tabatha Fireman (c). **Dreamstime.com:** Constantin Sava (bl). **165 Alamy Images:** Hemis (br). **Corbis:** Godong / Julian Kumar (tr). **Dreamstime.com:** F9photos (cr); Teptong (crb). **Getty Images:** Philippe Lissac (tc). **172 Alamy Images:** GL Archive (tr); The Art Archive (cb). **Corbis:** Bettmann (cl, cr); Oscar White (cla); The Gallery Collection (crb). **Dorling Kindersley:** Philip Keith Private Collection / Judith Miller (br). Getty Images: De Agostini (cra, cra/Gainsborough); Stringer / Powell (tc). **172–173 123RF.com. 173 Corbis:** (cl, cr, cb); Contemporary African Art Collection Limited (clb). **Getty Images:** AFP (bc); (tl, tr); (cla). **174 Corbis:** In Pictures / Barry Lewis (br). **175 Corbis:** JAI / Michele Falzone (cra). **Dorling Kindersley:** Rough Guides (bc); Surya Sankash Sarangi (c). **176 Dorling Kindersley:** Alex Robinson (br). **177 Corbis:** Jose Fuste Raga (bc). **178–179 Dreamstime.com:** Luminis (background image). **179 Dreamstime.com:** Mathayward (bl). **180 Alamy Images:** Aerial Archives (cl). **Getty Images:** (ca). **180–181 Getty Images:** AFP (cb); (ca). **181 Corbis:** Arcaid / John Gollings (br). **Getty Images:** (ca). **182 Corbis:** GT Images / George Tiedemann (tr); Icon SMI / Jeff Vest (br). **182–183 Dreamstime.com:** Eugenesergeev (tyre tracks on the map). **183 Getty Images:** (tl, tc, cr, bc). **Dreamstime.com:** Marco Canoniero (tr). **184 Alamy Images:** David Wall (tr). **Dreamstime.com:** Anthony Aneese Totah Jr (c). **Getty Images:** AFP (cl). **185 Alamy Images:** G.P.Bowater (tr); Philip Sayer (tc). Getty Images: AFP (br)

All other images © Dorling Kindersley
For further information see: www.dkimages.com